THE PRINCIPLE OF RELATIVITY

THE PRINCIPLE OF RELATIVITY

by

E. CUNNINGHAM, M.A.

FELLOW AND LECTURER OF ST JOHN'S COLLEGE, CAMBRIDGE

Cambridge:

at the University Press

1914

CAMBRIDGE UNIVERSITY PRESS
Cambridge, New York, Melbourne, Madrid, Cape Town, Singapore,
São Paulo, Delhi, Dubai, Tokyo, Mexico City

Cambridge University Press
The Edinburgh Building, Cambridge CB2 8RU, UK

Published in the United States of America by Cambridge University Press, New York

www.cambridge.org
Information on this title: www.cambridge.org/9780521156868

First published 1914
First paperback printing 2010

A catalogue record for this publication is available from the British Library

ISBN 978-0-521-15686-8 Paperback

PREFACE

THE controversial note which has been characteristic of discussions in respect of the Principle of Relativity has prevented the significance of the principle from being seen in its proper proportions and in its relation to general physical theory. On the one hand, there have been those who have magnified its importance, and assigned to it an unduly revolutionary power, while on the other hand, there are those who have scoffed at it as fantastic and reared on the most slender of physical bases. It has therefore seemed desirable in the first part of this book to outline the way in which the Principle of Relativity grew out of electrical theory, so that it might be made clear that there is a real place for it as a hypothesis supplementary to and independent of electrical theory owing to the limitations to which that theory is subject.

It is hoped that by drawing a clear distinction between the 'mode of measurement,' and the 'nature' of space and time, the author will escape from the charge of venturing unduly upon debatable metaphysical questions.

In the Second Part an attempt has been made to present in a simple form the more attractive of the two mathematical methods devised by Minkowski for the purpose of putting in evidence the relative nature of electrical and other phenomena.

The Third Part seeks to indicate some of the most fundamental points in which mechanical theory needs modification if the principle is accepted as universal. It has not been thought advisable to give an account of the purely formal and rather academic developments of special branches of mechanics such as hydromechanics, and elasticity, as these might tend to divert attention from the bearing of the principle on what are generally classified as the fundamental concepts. Some account of these is given by M. Laue in the second edition of his book, *Das Relativitätsprinzip*, Braunschweig, 1913. No attempt has been made to present the highly speculative attempt of Einstein at a generalization of the principle in connection with a physical theory of gravitation.

Throughout the intention has been as far as possible to consider those aspects of the principle which bear directly on practical physical questions. The mathematical part has been compressed to as small a compass as is consistent with furnishing sufficient apparatus for a systematic consideration of the problems suggested.

In the preparation of the book the author has received great help from Mr H. R. Hassé, who read the whole of the manuscript and made many suggestions for its improvement, besides reading the proofs of nearly the whole work. Mr R. W. James has also given valuable assistance in reading both manuscript and proofs. Especially would the author wish to acknowledge his debt to Sir Joseph Larmor, both personally and through his published works, for much stimulus and encouragement in the study of theoretical physics, and for valuable criticism of the earlier part of this book.

To the staff of the Cambridge University Press for care and courtesy in the work of printing the author is most grateful.

E. C.

CAMBRIDGE,
June 1914.

SUMMARY OF CONTENTS

CHAPTER III

THE ELECTRON THEORY

CHAPTER IV

CORRELATION OF STATIONARY AND MOVING SYSTEMS

CHAPTER V

EINSTEIN ON THE RELATIVITY OF SPACE AND TIME

CHAPTER VI

THE KINEMATICS OF EINSTEIN

CHAPTER VII

THE ELECTRON THEORY OF MATTER

PART II

MINKOWSKI'S FOUR-DIMENSION WORLD

CHAPTER VIII

MINKOWSKI'S FOUR-DIMENSION CALCULUS

CHAPTER IX

THE FIELD EQUATIONS OF THE ELECTRON THEORY IN MINKOWSKI'S FORM

CHAPTER X

MINKOWSKI'S ELECTRODYNAMICS OF MOVING BODIES

PART III

THE TRANSITION TO MECHANICAL THEORY

CHAPTER XI

THE DYNAMICS OF THE ELECTRON

CHAPTER XII

RELATIVITY AND DYNAMICAL THEORY

CHAPTER XIII

THE DYNAMICS OF A PARTICLE

CHAPTER XIV

THE DYNAMICS OF CONTINUOUS MATERIAL MEDIA

CHAPTER XV

RELATIVITY AND AN OBJECTIVE AETHER

CHAPTER XVI

RELATIVITY AND PROBABILITY

CHAPTER XVII

CONCLUDING REMARKS

PART I

THE PRINCIPLE OF RELATIVITY IN RELATION TO GENERAL PHYSICAL THEORY

CHAPTER I

INTRODUCTION

1. **The Relativity of the Newtonian Dynamics.**

It is a commonplace observation in respect of Newtonian dynamics that although the fundamental laws assume at the outset an absolute frame of reference, yet they are not sufficient to determine *uniquely* what that frame of reference must be; or in other words, although to every moving point a velocity is assigned, yet the laws of dynamics as stated by Newton are not sufficient to determine a velocity which, more than any other arbitrary velocity, can be said to be *the* velocity of any particular point. Put more definitely, it is known that if any set of axes in space can be specified relative to which the laws of dynamics are satisfied by a system of bodies, then any other set of axes which moves continually with a constant velocity of translation and with no rotation relatively to the former set, is also a valid framework for the dynamics of the same system of bodies.

It is important to distinguish this *dynamical relativity* from any philosophic dogma as to the *a priori* impossibility of the mind conceiving of an absolute position or motion in space. For such a proposition would be of much wider content than that stated above. It would imply not only that the velocity of a point is an undetermined quantity but also that its acceleration cannot be determined. Further there is nothing to exclude from the scope of such a supposed philosophic truth the impossibility of defining uniquely the velocity of rotation of a body. In fact apart from physical phenomena, the mind can no more and no less conceive of an absolutely fixed direction than of an absolutely fixed position.

In the Newtonian description of the modes of motion of bodies however, absolute standards of position and direction are at the outset assumed, not as a philosophic doctrine, but as a means of co-ordinating the phenomena observed as simply as possible. The justification for the assumption lies in the fact that the co-ordination is possible, and the laws so framed contain within themselves the definition of the terms 'absolute position' and 'direction,' and they also *define*, though not uniquely, 'the frame of reference.'

A *straight line* from the point of view of Newtonian dynamics is nothing more than the path of a particle which moves under the influence of no external agencies*, or again it is the locus of the positions at any instant of a number of particles which were projected in a common direction with different velocities from a given point at some previous moment.

2. The Space and Time of dynamics.

But more than this, the laws of dynamics contain statements as to distances and intervals of time. In the scheme as finally drawn up it appears as if the measurements of distance and of time were almost intuitive processes, the method to be adopted being undefined. It appears almost as if space were marked out permanently, and as if there were an absolute clock to

* The differential equations

$$\frac{d^2x}{dt^2}=0, \quad \frac{d^2y}{dt^2}=0, \quad \frac{d^2z}{dt^2}=0$$

lead at once to the equations

$$x=a_1t+b_1, \quad y=a_2t+b_2, \quad z=a_3t+b_3,$$

out of which all the descriptive geometry of the straight line and plane follow immediately.

which all time is referred. But on examination it is clear that our intuitive conception of distance is not sufficiently definite, and that all possible clocks are too dependent on dynamical mechanism to be of use as primary standards of time.

Thus, as in the case of position and direction, the definitions of the 'measures of space and time' in dynamics are partly contained within the laws of motion. Those laws at once express a uniformity in the phenomena of motion, and define a scheme of quantities in terms of which that uniformity may be simply expressed, but, as has been said, the scheme so defined is by no means unique. In the light of the principles of conservation of linear and angular momentum of a system of particles, all that can be stated is that we have criteria of certain fixed directions, namely the directions of the resultant linear and angular momenta, and also a criterion of uniform motion in a fixed direction, namely the motion of the centre of mass, of the system.

It was said above that the definitions of the measures of space and time are *partly* contained in the laws of dynamics. The two concepts are not in fact completely defined by the dynamics of separate particles. If we had an independent measure of time, then uniform motion would give a means of passing to a measure of space. Conversely if we had an independent measure of space, then the measure of time could be obtained. The clock would be a free particle traversing a graduated straight line.

But if within the term 'dynamics' we include the phenomena of a rigid body, then both measures are defined—subject to a choice of units. For in this branch of dynamics the ideal rigid body is conceived as having a definite size; the distance between two points of it is conceived as always defining the same interval of distance, whether the body is moving or not. It may be looked upon as a means of measuring space. Or again if, avoiding the usual somewhat artificial deductions of the equations of motion of rigid bodies from those of separate particles, we take those equations for granted as the appropriate extension of Newton's equations, then we have an ideal clock in a body rotating about an axis of symmetry, and hence, by means of the motion of free particles, a method of graduating space. But it is implied in our scheme of dynamics that if space were so graduated then the distance between two given points of any rigid body would occupy the same interval wherever it was placed, and whatever were the free particle which was defined to be at rest.

3. There is then a **Newtonian principle of relativity** which may be stated thus:

It is impossible by means of any dynamical phenomena to ascertain absolutely the velocity of any material particle, but the relative velocity of any point with respect to any other is a uniquely determinable quantity.

It is worth noticing at once that in considering the same dynamical system from two different frames of reference moving relatively to one another with a certain uniform velocity of translation, there are certain quantities that have the same value in the two cases. The 'mass' of a particle, its 'acceleration,' the 'force' acting on a particle, the 'distance between two marked points of a rigid body,' and the 'measure of an interval of time' are the most fundamental. On the other hand the velocity, momentum, and energy vary with the frame of reference.

It will be important in the ensuing discussion to emphasize the invariant quantities, but the main outlook is towards the *invariant relations* between the quantities which are not themselves invariant.

For example, if the vector **u*** is the velocity of a moving particle relative to a certain Newtonian frame of reference we have

$$m\frac{d\mathbf{u}}{dt}=\mathbf{P},$$

where **P** is the force acting.

* Throughout the book vector quantities are represented by Clarendon type, and $(\mathbf{uv})$ represents the scalar product of $\mathbf{u}$ and $\mathbf{v}$, i.e. $|\mathbf{u}||\mathbf{v}|\cos\theta$, where $|\mathbf{u}|$ is the magnitude of $\mathbf{u}$ and θ is the angle between $\mathbf{u}$ and $\mathbf{v}$.

If $\mathbf{u}'$ is the velocity referred to a new frame of reference whose velocity relative to the first is $\mathbf{v}$

$$\mathbf{u}' = \mathbf{u} - \mathbf{v},$$

so that $\mathbf{v}$ being constant

$$m\frac{d\mathbf{u}'}{dt} = \mathbf{P}.$$

Here m and $\mathbf{P}$ are *invariant quantities* and the equation is an *invariant relation.* It is the relation which is all important.

Again the equation of energy in the first system is

$$\frac{d}{dt}(\tfrac{1}{2}m\mathbf{u}^2) = (\mathbf{Pu}),$$

and in the second

$$\frac{d}{dt}(\tfrac{1}{2}m\mathbf{u}'^2) = (\mathbf{Pu}');$$

the latter can be obtained from the former by means of the equation of momentum.

But conversely *if we assume the equation of energy to be an invariant relation, the equation of momentum may be deduced.*

Thus

$$\frac{d}{dt}(\tfrac{1}{2}m\mathbf{u}'^2) = \frac{d}{dt}(\tfrac{1}{2}m\mathbf{u}^2) - \frac{d}{dt}(m\mathbf{uv}) + \frac{d}{dt}(\tfrac{1}{2}m\mathbf{v}^2),$$

or since $\mathbf{v}$ is constant

$$(\mathbf{Pu}') = (\mathbf{Pu}) - \left(\mathbf{v}\frac{d}{dt}m\mathbf{u}\right),$$

or

$$(\mathbf{vP}) = \left(\mathbf{v}\frac{d}{dt}m\mathbf{u}\right).$$

If this is true whatever the value of the relative velocity $\mathbf{v}$ of the two frames of reference, we must have

$$\mathbf{P} = \frac{d}{dt}m\mathbf{u}.$$

4. Remembering that a force is not a primary physical quantity*, it will be as well to refer to the more fundamental physical results of the conservation of energy and momentum for a system of particles under no forces but their own mutual actions.

* As commonly presented in present day teaching—but see § 5 and appendix, pp. 9 and 10.

From the equations

$$\Sigma \tfrac{1}{2} m \mathbf{u}^2 + V = \text{const.},$$
$$\Sigma m \mathbf{u} = \text{const.},$$

in which the potential energy of the configuration is V, it follows that

$$\Sigma \tfrac{1}{2} m (\mathbf{u} - \mathbf{v})^2 + V = \text{const.},$$

where $\mathbf{v}$ is any constant vector. Thus if it be taken that V is a function of the *relative* positions of the particles (which are invariant quantities), so that V is an invariant, it follows that the sum of the kinetic and potential energies is constant whatever be the velocity.

In the same way if it be assumed that V is independent of $\mathbf{v}$ and that the constancy of the sum of the energies is an invariant relation, it follows that $\Sigma m\mathbf{u}$ is constant; or, in other words—*the principle of the conservation of linear momentum follows from the principle of energy combined with the principle of relativity.*

In this statement of course the mass m becomes merely a constant in the equation of energy—its definition is included in the assumption of the existence of such an equation, and the familiar definition in terms of accelerations takes a derived position.

5. It is striking that the invariant quantities *mass* and *force* are precisely those which have, in the point of view represented by Mach (*Science of Mechanics*) and Karl Pearson (*Grammar of Science*) among others, been relegated from the position of primary concepts to that of mathematical numbers characteristic of certain uniformities and relations between the accelerations of bodies. But it is at least capable of being argued* that even if the means of measuring them is lacking, qualitatively they are matters of apprehension as direct as time and space; and, as for measurement, the ordinary means applied to the latter depend upon the properties of mass, force, and rigidity. It is as easy to imagine the reproduction of the physical

* See quotations in the appendix to this chapter.

conditions giving rise to what we call a *force*, as to conceive of an ideally periodic phenomenon which will serve as a clock.

In passing from the 'dynamics of experiment,' in which forces are looked upon as prior to motion, as for example in the projection of a bullet from a rifle, or the throwing of a ball by muscular effort, to the 'dynamics of theory,' we are, as in nearly all logical schemes, extracting from the complex of experience of motion in space ideal conceptions of abstract space and time merely as a foundation or background for an ordered description of the permanent relations between different factors in that experience. When we have obtained these two conceptions by an analytic process from the directly perceived facts of motion, we can by a synthetic process build up from these materials an ideal conception of motion which records the facts of experience in a form amenable to mathematical treatment.

But as the essence of a picture is not in the canvas, but in the painting thereon as a record and interpretation of some part or aspect of human experience; so the value of dynamical theory is not in its reducing 'force' and 'mass' to abstractions by making 'space' and 'time' the prime concepts, but in its giving a fuller and more precise significance to those fundamental elements of experience by making a picture of them which is complete in itself*.

6. Statement of the General Principle of Relativity.

The Principle of Relativity which is the subject of this work holds the same place in the physical thought of to-day, that the foregoing principle of dynamical relativity held in the time when the laws of dynamics were considered as ultimate and all-embracing. It consists in the general hypothesis, based on a certain amount of experimental evidence, that the *problem of determining in a physical sense the absolute velocity of a body is one that can no more be solved uniquely by the help of optical and electrical phenomena, than it could by means of dynamical observations.*

* Cf. Larmor, *Aether and Matter*, Appendix B, especially § 3, pp. 271–3. See also pp. 9, 10.

If we speak of a 'fixed aether' as the background of electrical activity, it is the hypothesis that *the velocity of any piece of matter relative to the aether is unknowable.*

To put it more precisely, it states *that we neither have nor expect to have any experimental evidence of the uniqueness of the framework which we call 'the aether,' but that if there is one, there is an infinite number of such frames of reference, any one of which has, relative to any other, a uniform velocity of translation without rotation, this velocity being of arbitrary magnitude.*

In this statement, it will be seen, the old philosophic difficulty as to *absolute direction* or *angular velocity* remains. The domain of the principle is co-extensive with the relativity which follows from the laws of dynamics. But as in that case, the assertion of the principle is not a metaphysical dogma. It is an *empirical* principle, suggested by an observed group of facts, namely the failure of experimental devices for determining the velocity of the earth relative to the luminiferous aether, and would make it a *criterion* of theories of matter that they should give an account of this failure, and it suggests modifications where the theory is insufficient to do so. But like all physical principles, it is to be probed by further experience.

It will be seen that it involves a reconsideration of many old-established preconceptions. It emphasizes very strongly what has been said as to the derivative nature of *metrical* space and time; though, of course, as a physical principle it has nothing to say against the reality of the perception of spatial extent and temporal duration. It is in fact completely dependent upon such perception.

It will be found that many commonly accepted terms such as 'simultaneity,' 'electric force,' 'aether' are incapable of unique definition or specification, and to that extent lose that reality which they seemed to possess. A criterion of *objective* or *physical* reality is strongly suggested in the requirement of *uniqueness of definition* and *invariance of magnitude* whatever the frame of reference chosen out of an infinite number that are possible.

Thus, for example, in the light of what has been said, the metrical absolute space of Newton is not unique. On the other hand, in *force* and *mass* we are dealing with quantities which the relations perceived between phenomena do not leave ambiguous*, so that, in this sense, we may say that they have a *physical* significance such as we cannot attribute to velocity, or energy.

Thus we are reminded that mechanics was in its origin a calculus of *forces*, that Galileo's great achievement was in the clear statement of the property of *mass* or inertia as a permanent property of matter, and that only at a later date was the attempt made to frame a system of Mechanics in which these quantities took a secondary place, while motion relative to a conceptual and undefined† framework was given the pre-eminence as the basis of the science.

APPENDIX

It may be worth while for the sake of emphasizing the point of view to place side by side quotations representing different schools of thought.

Karl Pearson—*Grammar of Science*, 3rd edition, p. 332.

"The definition of force we have reached is a perfectly intelligible one; it is completely freed from any notion of matter as the moving thing, or from any notion of a metaphysical cause of motion....Force is an arbitrary conceptual measure of motion without any perceptual equivalent."

Contrast with this

Larmor—*Aether and Matter*, p. 272.

"To say, as is sometimes done, that force is a mere figment of the imagination which is useful to describe the motional changes that are going on around us in Nature, is to assume a scientific attitude that is

* Subject of course to the restriction of the phenomena in question to a certain limited field.

† That is, undefined apart from the relations to be subsequently developed.

appropriate for an intelligence that surveys the totality of things: but a finite intellect, engaged in spelling out the large-scale permanences of relations in material phenomena, is not cognizant of the bulk of these ultimate motions at all, and must supply the defect by the best apparatus of representation of the regular part of their effects that is in his power. When a person measures the steady pull of his arm by the extension of a spring, where or what, for example, are the motions of which the pull is only a mode of representation? The only way of gradually acquiring knowledge as to what they are, is to develope and make use of all the exact concepts that examination of the phenomena suggests to the mind. And in any case it is not the motions that are the essential factors, so much as the permanent entities of which the motions merely produce rearrangement."

Also p. 278.

"...it would be a misfortune to banish the idea of force even if we could. Any definition that would merely make it a subjective cause of motion is incomplete: the concept is required for the expression of properties of permanent groupings of natural phenomena, and in that sense is as much objective as anything else. When we hang up a given weight on a spring-balance the extension of the balance is always the same, subject to permanence of locality and other assignable conditions, and whenever we see the spring so extended we infer at once that it is supporting an equal weight or else doing something equivalent: we say that it is exerting a certain definite force. It would of course be useless to introduce this conception of force if the uniformity of the course of Nature did not hold to the extent here described. But as it does hold, the force is the concept that allows us to eliminate the consideration of the complex of changes of molecular states and motions that is involved in the extension of the spring, of which we know nothing except that they are for our purposes the same in each case."

And p. 279.

"On the other hand, if the idea of force had not been supplied to us ready formed, through our muscular sense, we can conceive that the science of Mechanics must have begun with the dynamics of molecular systems, and the forces between permanent finite bodies would have been discovered and defined as new physical conceptions simplifying the theoretical discussions and related to the degree of permanence of the systems: the conception of the potential in electrostatics is actually one of this kind: so is that of temperature, which also was early developed becaus our sense of heat supplied it ready formed."

CHAPTER II

HISTORICAL

1. **The Early Development of the Concept of the Aether*.**

The foregoing chapter has been prefixed in order to prepare the way for the ideas which are to be developed in the region of electromagnetic phenomena. The phenomenon of aberration discovered by Bradley in 1727, though it revealed something of the relative motion of the earth and the stars, went no way towards clearing up the difficulty of determining an absolute velocity of bodies through space. It was quite simply explained on an emission theory of light, which makes the variation in the apparent position of the star depend on the variation in the velocity of the earth relative to a Newtonian frame of reference. Such a frame of reference is required for the purposes of explanation; but when that is completed it is only relative velocities that appear in the result, and in accordance with Newtonian theory these relative velocities or changes of velocity have a unique and defined value.

But with the rise of the undulatory theory of light and the conception of a luminiferous aether came difficulties and questions. What is this aether, and what its relation to matter? How is it influenced by the motion of the earth and other bodies through it?

The natural conception is of a fluid of some kind which may or may not penetrate into the interstices of matter. Some thinkers were for excluding it entirely from the space

* **An exhaustive account of this part of the subject up to the year 1900 may be found in Larmor's *Aether and Matter*, 1900, especially in the historical survey which forms the first chapter of that work.**

occupied by material bodies, a view which necessitated it being moved as they moved. Others adopted a modification of this view, allowing that the aether might penetrate and pass through the interstices of matter, but that the motion of the matter might exercise a dragging influence on it as it moved through it.

The extreme position on one hand is that the aether is comparable with a fluid to which matter is impervious, but which is pushed and dragged at the bounding surface of any material body, in such a way that at the boundary the velocity of the aether is equal to the velocity of the matter. This was the view taken by Arago, Cauchy and Stokes. The main objection to it lay in the difficulty of explaining aberration from this point of view, but Stokes shewed that this was not insuperable.

On the other hand lies the extreme position that the aether is stagnant or immobile, that the passage of matter through it produces no disturbance of it as a whole. Although this view was hinted at by early writers, it only came gradually into general acceptance, and this probably because the early conception of the aether was that of a particular form of matter, which could not coexist with any other form in the same place.

2. Arago's experiment and Fresnel's convection-coefficient. 1818.

One of the earliest experiments bearing on the solution of the problem was conceived and carried out by Arago*. His idea was that since the deviation of a ray of light by a prism depends on the ratio of the velocity of light in space to its velocity through the material of the prism, the motion of the prism through space, since it affects the relative velocity with which the light meets the refracting surface, will also affect the amount of the deviation. He calculated that the effect produced would be possibly as much as a change of a minute of arc, and this was a quantity well within the reach of observation. But on trying

* See Fresnel's letter to Arago, *Annales de Chimie*, 1818, quoted by Larmor in *Aether and Matter*, p. 320.

the experiment he found that the change was not to be perceived, and the conclusion he drew was that the aether in the neighbourhood of the earth was dragged along with the earth, so that the velocity with which the ray met the prism was simply the constant velocity relative to the aether.

The result of his experiment he communicated to Fresnel who accepted the result but not the explanation. He pointed out that it was possible to maintain the conception of a stationary luminiferous medium, by supposing that the velocity of light in the prism, relative to it, was affected by the motion through the aether. Arago had not contemplated this possibility. Fresnel was able to shew that there would be no change in the deviation if, the velocities of light in free space and in the prism at rest being c and c', the corresponding velocities relative to the prism when moving with velocity v in the direction of the light were $c - v$ and $c' - v/\mu^2$, μ being the ordinary index of refraction c/c'; or, in other words, if the absolute velocity of light in the moving prism were $c' + v(1 - \mu^{-2})$. The conception suggested by Fresnel is that the aether permeates the moving matter and is partially convected by it, being absorbed at the front surface and emitted again in the rear of it*.

3. **Fizeau's experiment. 1851.**

The truth of this explanation of Fresnel's was one capable of experimental verification. Fizeau† devised an experiment in which a beam of light was divided into two parts which traversed two parallel tubes filled with water which could be set in motion with a measurable velocity. In the figure *AB*, *CD* represent the tubes. One portion of the beam of light starting from *S* travels along the path *SLMT* being

* It is easily seen that if a thick plate advances at right angles to its surfaces absorbing a fluid through which it moves and the fluid is conceived to be condensed within the plate to a density k times as great as its original density, and is emitted again behind, the velocity of the condensed fluid within the plate is $(1 - 1/k)$ times the velocity of the plate, the fluid outside being always at rest.

† *Comptes Rendus*, 33 (1851), p. 349; *Ann. d. Phys.* 1853, p. 377.

reflected at L and M, while the other portion travels along the same path in the opposite direction. The two parts are then reunited, and interference is observed.

The water in the tubes is then caused to circulate, moving in opposite directions along AB and CD, so that it is moving with one part of the beam of light, and against the other. A shifting of the interference fringes is at once observed which is proportional to the velocity of the water.

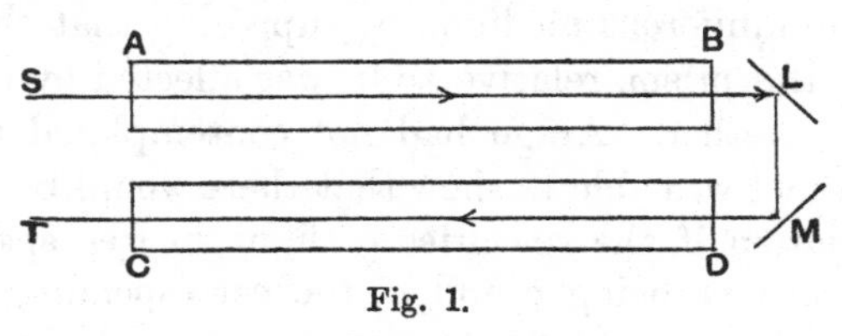

Fig. 1.

Supposing that the velocity of light through the moving water is $c' + kv$, where c' is the velocity when the velocity v is zero, the times taken by the two parts of the beam to traverse the total path l in the water are respectively

$$\frac{l}{c' + kv} \quad \text{and} \quad \frac{l}{c' - kv},$$

the velocity v being in opposite senses for the two parts. Thus the retardation of the one beam relative to the other is

$$\frac{l}{c' - kv} - \frac{l}{c' + kv} = \frac{2lkv}{c'^2},$$

neglecting $(v/c')^2$.

On measuring the displacement of the interference fringes, Fizeau found that Fresnel's value for the convection-coefficient k, viz. $1 - \mu^{-2}$, fitted his results quite well. The conclusion was confirmed later by Michelson and Morley who repeated the experiment with modern refinements*.

4. The significance of Fresnel's convection-coefficient, and of Fizeau's verification of it.

The suggestion of Fresnel, and the experiment confirming it, exercised a permanent influence on the conception of the

* See below, Chapter VI, pp. 63–4.

nature of the means by which light is propagated. The velocity of light relative to a material medium is definitely shewn not to be a constant depending on the nature of the medium alone, but to be also dependent on the velocity of the medium through space, or rather through the aether which was conceived by Fresnel to be at rest in space.

The effect being a first order one, that is being proportional to the velocity, the experiment threw no light on the velocity of the earth itself, since the part of the effect due to this velocity remains practically constant, the variable part being proportional to the velocity of the water in the tubes relative to the earth. But henceforth it was clear that in respect of its optical properties a material medium was in some sense modified by its motion.

When at a later date light was identified as an electromagnetic disturbance this became one deciding factor as between Hertz' theory of electromagnetic phenomena in moving bodies, and that of Maxwell and Lorentz. Hertz' idea of a moving body was just an extension of the Newtonian idea of a rigid body in motion. The properties of the body were entirely unchanged when it was set in motion, just as the dynamical mass of a body was unaltered; and among other implications was the complete convection of light, that is the invariance of the velocity of light relative to the body. In view of Fizeau's experiment it was not possible to maintain this theory, and other experiments soon shewed it to be lacking in other respects.

Fresnel's suggestion of the interpenetration of aether and matter therefore assumed greater prominence and became the basis of the later theory developed by Lorentz and by Larmor.

5. The Michelson-Morley experiment*.

The result of the experiment of Fizeau, together with the fact of aberration, gradually told in favour of the notion of a stagnant aether.

In 1887 an experiment was devised and carried out by

* *Phil. Mag.* Dec. 1887.

A. A. Michelson and afterwards with greater refinement by E. W. Morley which seemed to promise a possibility of determining the velocity of the earth relative to the aether, assuming it to be of the same order of magnitude as that of the earth relative to the sun.

Fizeau's experiment, as has been pointed out, being a first order experiment, could only reveal the influence of velocity relative to the earth. In the Michelson and Morley experiment it was proposed to seek for an effect depending on the square of the velocity, and to avoid any question concerning the internal constitution of moving matter. The arrangement was as follows, the figure shewing the path of the light *relative to the moving apparatus.*

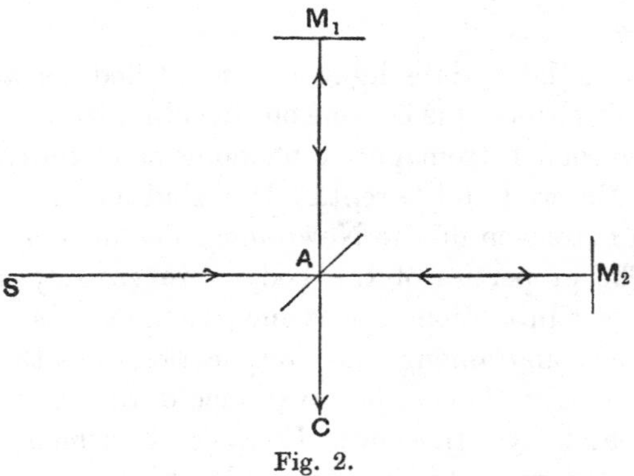

Fig. 2.

A beam of light from a source S was divided by partial reflection at a plate of glass A into two portions travelling along the paths AM_1, AM_2. These two portions were reflected back by mirrors M_1 and M_2 and on striking the plate again, a portion of the beam from M_1 is transmitted and brought to interference with a portion of the beam from M_2 reflected along AC. The whole apparatus could be rotated into any position desired.

The theory of the experiment.

Suppose first that AM_2 is in the direction of the earth's velocity, relative to the aether, which will be called v. Then since $(c-v)$ is the velocity of light relative to the apparatus on the forward journey and $(c+v)$ is the relative velocity on the return journey, the time taken by the light to travel from A to M_2 and back is

$$\frac{l_2}{c-v}+\frac{l_2}{c+v}=\frac{2l_2c}{c^2-v^2},$$

where l_2 is the distance AM_2.

In considering the time taken to go to M_1 and back we have to take a ray whose velocity *relative to the apparatus* is along AM_1, which is assumed perpendicular to AM_2. The relative velocity along this line will thus be $(c^2-v^2)^{\frac{1}{2}}$, and the time taken is therefore

$$\frac{2l_1}{(c^2-v^2)^{\frac{1}{2}}},$$

where l_1 is the distance AM_1.

Thus the retardation in time of the former beam relative to the latter when they are brought together again is

$$2\left\{\frac{l_2c}{c^2-v^2}-\frac{l_1}{(c^2-v^2)^{\frac{1}{2}}}\right\}.$$

If the apparatus is now rotated so that AM_1 comes into the direction of the light, l_1 and l_2 change places, and the retardation is altered to

$$2\left\{\frac{l_2}{(c^2-v^2)^{\frac{1}{2}}}-\frac{l_1c}{c^2-v^2}\right\}.$$

Thus the change in the retardation is

$$2(l_2+l_1)\left\{\frac{c}{c^2-v^2}-\frac{1}{(c^2-v^2)^{\frac{1}{2}}}\right\},$$

which, neglecting powers of (v/c) higher than the second, is equal to

$$(l_2+l_1)\,v^2/c^3.$$

Alternative explanation.

The following explanation, alternative to that given above, and fig. 3 which illustrates the absolute path of the rays which interfere, may perhaps make the theory of the experiment a little clearer.

At a certain moment t_3 let A_3 be the position of the moving plate which divides the beam, and let us consider the paths of the two parts of the beam which are reunited at that moment. Let A_1 be the position of the plate at time t_1 when that element of disturbance leaves A_1 which, travelling by the path $A_1M_1A_3$, arrives at A_3 at time t_3.

Then the time t_1 is found thus:

$$A_1A_3 = v(t_3 - t_1),$$

$$c(t_3 - t_1) = A_1M_1 + M_1A_3$$

$$= 2\sqrt{l_1^2 + (A_1A_3/2)^2}$$

$$= \sqrt{4l_1^2 + v^2(t_3 - t_1)^2},$$

giving $$(c^2 - v^2)(t_3 - t_1)^2 = 4l_1^2,$$

or $$t_3 - t_1 = \frac{2l_1}{(c^2 - v^2)^{\frac{1}{2}}}.$$

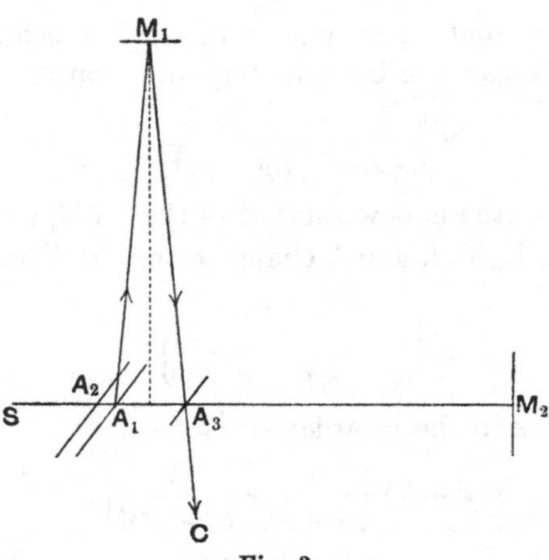

Fig. 3.

In the same way let A_2 be the position of the plate at the moment t_2 of the emission of the element of disturbance which, travelling by the path $A_2M_2A_3$, arrives at A_3 at time t_3.

Then $$A_3A_2 = v(t_3 - t_2),$$

$$A_2M_2 = c(\tau - t_2),$$

$$M_2A_3 = c(t_3 - \tau),$$

where τ is the moment of reflection.

Also l_2 is the distance of M_2 from the position of the plate at time τ.

Thus
$$l_2 = A_2M_2 - v(\tau - t_2)$$
$$= (c - v)(\tau - t_2),$$
and similarly
$$l_2 = (c + v)(t_3 - \tau),$$
giving
$$\frac{l_2}{c - v} + \frac{l_2}{c + v} = t_3 - t_2,$$
or
$$t_3 - t_2 = \frac{2l_2c}{c^2 - v^2}.$$

Thus the disturbances which are united at time t_3 at A_3 do not leave the plate A simultaneously, but at instants separated by an interval

$$t_2 - t_1 = \frac{2l_1}{(c^2 - v^2)^{\frac{1}{2}}} - \frac{2l_2c}{c^2 - v^2},$$

and therefore will differ in phase by this amount, exactly as obtained above. The question to be answered by the experiment is whether this difference of phase will be altered when the apparatus is turned round.

Looking at the theory from this point of view it becomes necessary to consider whether the directions of the reunited parts of the beam will necessarily be the same. This involves the question of the reflection of a ray of light at a moving mirror. Let us consider this by means of Huygens' Principle.

Let AB be a reflector placed at any angle α with SS' moving with velocity v in the direction SS'. Let AX be a plane wave-front incident on the mirror at A at a certain instant; at a small time τ later, the unreflected portion of this will have advanced a distance $c\tau$ being now part of the line $X'N$. In the same time the reflector has advanced a distance $AA' = v\tau$; so that, drawing $A'B'$ parallel to AB to meet $X'N$ in C, C is now the point of incidence of the wave-front on the reflector. Thus the reflection is exactly that which would take place at a fixed mirror in the position AC.

In the same way, considering a ray incident on the other side of AB, moving in the opposite direction, we find that $A'C'$ is the direction of the equivalent fixed mirror.

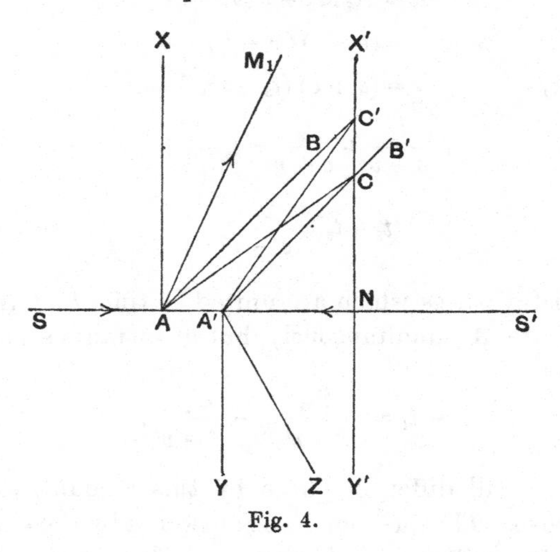

Fig. 4.

Thus if the mirror is set at an angle of 45° with AS' the reflected part of the beam at the first incidence will be AM_1 where the angle

$$X\hat{A}M_1 = 2C'AC,$$

and the direction of the transmitted ray after its final reflection will be $A'Z$ where

$$YA'Z = 2C'A'C.$$

The condition that the reunited rays shall be parallel is that

$$XAM_1 = YA'Z,$$

that is that

$$CAC' = C'A'C,$$

and this is satisfied since the right-angled triangle ANC' is isosceles.

*The result of the experiment**.

Interpreted according to this calculation, if the velocity of the earth relative to the aether was as much as a quarter of the

* Michelson and Morley, *American Journal of Science*, 34 (1887).

velocity of the earth in its orbit, there should have been a displacement of the fringes produced by the interference of the two beams of such magnitude that it could not have escaped observation with the apparatus used. But no trace of such a displacement was found.

Morley and Miller* repeating the experiment in 1905 with still greater refinement, also came to the conclusion that there was no displacement, though they could have observed one due to one-tenth of the velocity of the earth.

6. The FitzGerald-Lorentz Contraction.

If we accept the above results, as we seem bound to, as shewing the theory which has been given to be inadequate, the hope of determining v by this means vanishes, and the difficulty remains of reconciling the null result with the hypothesis of an aether which is not convected along with the optical system by the earth.

FitzGerald† threw out a suggestion that *if the aether can percolate through matter, it may affect the apparatus and change its dimensions when it is rotated.* Such a suggestion had become feasible in view of the adoption of the electromagnetic theory of light, and the growing knowledge of the electrical relations of matter‡.

If such an effect is to nullify the change in the retardation, we must have, if l_1', l_2' are the changed lengths of AM_1, AM_2,

$$\frac{l_2'}{(c^2-v^2)^{\frac{1}{2}}} - \frac{l_2 c}{c^2-v^2} = \frac{l_1' c}{c^2-v^2} - \frac{l_1}{(c^2-v^2)^{\frac{1}{2}}},$$

and this is satisfied if

$$\frac{l_2}{l_2'} = \frac{l_1'}{l_1} = \left(1-\frac{v^2}{c^2}\right)^{\frac{1}{2}},$$

that is if either arm contracts in the ratio $1:\left(1-\frac{v^2}{c^2}\right)^{\frac{1}{2}}$ when

* E. W. Morley and D. C. Miller, *Phil. Mag.* 9 (1905).

† See Lodge, "Aberration Problems," *Phil. Trans.* 184 A (1893), p. 727, also Presidential Address to the British Association, 1913.

‡ E.g. Maxwell's law connecting the index of refraction with the dielectric constant.

turned into the direction of motion, as compared with its length when at right angles to this direction.

FitzGerald's suggestion was not carried further at the time, and it was left to Lorentz* to make it independently, shewing at the same time some plausible reason why a contraction of exactly this amount might be expected.

The theory to which Lorentz was led in his investigations into the optical behaviour of moving bodies is so fundamental to the present subject, and is, if not always in its original form, so generally accepted, that some account of it must now be given.

7. The null result independent of the material constituting the apparatus.

Before this is done, however, we may note a fact of great importance for the general significance of the Michelson-Morley experiment. As first carried out the whole apparatus was mounted on a sandstone block which floated in mercury. In the repetition by Morley and Miller, the distance between the mirrors was intentionally maintained by wooden rods.

Thus if the null effect is to be explained by FitzGerald's suggestion, the contraction of the right amount must automatically take place in two such different materials as sandstone and pine. It is difficult to think of this as a mere coincidence, so that we are naturally led to think that, if we accept the contraction hypothesis, whatever the explanation of it may be, it is one that is universal and inherent in the constitution of matter, even down to the structure of the chemical atom.

* *Versuch einer Theorie der elektrischen und optischen Erscheinungen in bewegten Körpern*, Leiden, 1895.

CHAPTER III

THE ELECTRON THEORY

1. The electron theory is based upon the fundamental assumption, that in the ultimate analysis all the electromagnetic properties of matter will be found to be due to the motions of electric charges, and that the equations of Maxwell for a stationary medium of specific inductive capacity unity and magnetic permeability unity are universally valid for determining the field set up by the flow of electricity through it. That is, using $\mathbf{e}$ and $\mathbf{h}$ for the electric and magnetic intensity vectors, by a proper choice of units the differential equations to determine $\mathbf{e}$ and $\mathbf{h}$, given the motion of the charges in the field, are

$$\frac{1}{c}\left(\frac{\partial \mathbf{e}}{\partial t} + \rho \mathbf{u}\right) = \operatorname{curl} \mathbf{h} \quad \text{(I)},$$

$$-\frac{1}{c}\frac{\partial \mathbf{h}}{\partial t} = \operatorname{curl} \mathbf{e} \quad \text{(II)},$$

$$\rho = \operatorname{div} \mathbf{e} \quad \text{(III)},$$

$$0 = \operatorname{div} \mathbf{h} \quad \text{(IV)}.$$

2. Conservation of electric charge, and atomic nature of charge.

These equations have an important implication which is characteristic of electron theory.

From (I) and (III) we obtain the equation

$$\frac{\partial \rho}{\partial t} + \operatorname{div}(\rho \mathbf{u}) = 0,$$

or
$$\frac{\partial}{\partial t}\iiint \rho\, dV = -\iint \rho(\mathbf{u}d\mathbf{S})^{*}$$

for any closed fixed volume, which implies that a decrease in the amount of electricity within a given volume can only alter by the actual passage of electricity across the boundary, so that no electricity is ever destroyed but only transferred.

It may be noted that it is implied in the equations as written that there is a unique velocity at any point. But they may be modified in such a way as to allow of two different kinds of electricity existing simultaneously at a given point and moving independently. The only modification required is the insertion of a sign of summation before the terms $\rho\mathbf{u}$ and ρ in equations (I) and (III).

It is assumed in the further development of the theory that the electricity exists in small nuclei, each having a definite charge, and that such nuclei cannot overlap or interpenetrate, and that the electromagnetic properties of matter are those of aggregates of such nuclei. But the atomic nature of electricity is an additional assumption not contained in equations (I)—(IV) nor explained by them.

3. The incompleteness of the field equations.

What is being attempted, in the construction of the electron theory, is to obtain a set of equations which shall be sufficient to determine the way in which all the electric charges and charged bodies will move, given the state of the whole field at a particular instant.

Now the equations (I)—(IV) are not sufficient to do this. Given the motion of the charges through all time, and the values of **e** and **h** for all space at a particular instant, the equations will determine the values of **e** and **h** for all time, but this is not the main problem of electrodynamics.

The vectors **e** and **h** are not themselves capable of direct

* Throughout the book dV will be used for an element of the volume of integration, and $d\mathbf{S}$ for an element of area.

observation, they only become apparent at points where charge exists. Lorentz therefore adds to these equations another, viz.

$$\mathbf{F} = \rho\,(\mathbf{e} + [\mathbf{uh}]/c)\quad\text{.....................(V)},$$

where $\mathbf{F}$ is said to be the *force per unit volume* acting on the charge or whatever carries the charge*.

This equation does not complete the set of equations in the sense mentioned above inasmuch as it introduces yet another variable $\mathbf{F}$.

If we suppose that the force acting on any element of charge is a quantity that can be directly measured, this equation attaches a meaning to the vectors $\mathbf{e}$ and $\mathbf{h}$, or rather to the single vector $(\mathbf{e} + [\mathbf{uh}]/c)$, at points where there is any charge, but for all other points $\mathbf{e}$ and $\mathbf{h}$ are so far nothing more than mathematical variables introduced for the purpose of calculation.

Some other fact then is still required to complete the set of equations, and it is vital to the present discussion to consider what is the extra assumption made in Lorentz's theory, and in other forms of the electron theory.

It would be possible of course to assume a relation between $\mathbf{F}$ and the acceleration of the charge at each point, so that, given ρ, $\mathbf{u}$, $\mathbf{e}$, $\mathbf{h}$ at a given instant, we could successively determine the increments $\delta\mathbf{e}$, $\delta\mathbf{h}$, $\delta\mathbf{u}$, $\delta\rho$ during a short interval of time δt, and so from instant to instant in theory calculate the field for all time. Such an assumption is suggested by Newtonian theory, but for various reasons has not been adopted.

4. The hypothesis of purely electromagnetic inertia.

The chief of these reasons is the rise of the belief in the atomic nature of electricity, demanding that at some point in the theory the electron or nucleus of electricity shall be treated as a whole and not in its infinitesimal elements.

This is done in two different ways associated separately with

* $[\mathbf{uh}]$ here represents the *vector-product* of $\mathbf{u}$ and $\mathbf{h}$, i.e. a vector at right angles to $\mathbf{u}$ and $\mathbf{h}$ of magnitude $|\mathbf{u}|\,|\mathbf{h}|\sin\theta$, θ being the angle between $\mathbf{u}$ and $\mathbf{h}$, the sense of $[\mathbf{uh}]$ being such that a right-handed rotation about it is a rotation from the direction of $\mathbf{u}$ to that of $\mathbf{h}$.

the names of Lorentz and Larmor. The former introduces a geometrical assumption as to the distribution of electricity within the nucleus, namely that the nucleus at rest is spherical, and in motion is merely altered by undergoing a FitzGerald contraction; thus the motion of every element of charge becomes dependent upon a single vector which represents the velocity of the electron as a whole. Lorentz then makes the assumption that *the total electromagnetic force* $\iiint \mathbf{F}\,dV$ *for the whole electron exactly balances the sum of any non-electromagnetic forces on it.* Thus a hypothesis of intrinsic inertia of electricity in its smallest elements is avoided at the expense of introducing an undefined mechanism for sustaining an arbitrarily selected distribution of charge within the electron.

5. In a theory starting from the properties of the free aether Larmor assumes the charge to exist only as definite identical structures so small that they are treated as points, assuming that in this limiting case questions relating to such structures will not enter, so that ρ is taken to be zero everywhere except at those points at which it becomes infinite.

In this form of the theory the fundamental equations become

$$\frac{1}{c}\frac{\partial \mathbf{e}}{\partial t} = \text{curl}\, \mathbf{h},$$

$$-\frac{1}{c}\frac{\partial \mathbf{h}}{\partial t} = \text{curl}\, \mathbf{e},$$

$$0 = \text{div}\, \mathbf{e},$$

$$0 = \text{div}\, \mathbf{h},$$

which appear to be complete*. But the kinematical assumption of Lorentz is only turned into the limitation that *only those solutions will be considered* which have singularities of a type corresponding to moving point charges of constant strength, the charge within any closed surface being measured by Gauss' Theorem.

* *Aether and Matter*, pp. 164–5.

Further when it is desired to connect the acceleration of the electron with the field in its neighbourhood, the structure of the electron has of course to be considered, or an undetermined constant depending on them has to be introduced*.

In this form of the theory the existence of other than electromagnetic forces is not contemplated, the field equations being supposed to determine the whole sequence of changes of the system.

The chief disadvantage of this form of the theory is that it appears in certain respects to have departed too far from the facts of experience. Thus for instance we can obtain a solution of the given equations by taking the ordinary electrostatic field of two point charges at rest. Hence if the given set of equations contains the whole of the laws of motion of charges it seems at first sight possible for two point charges to remain permanently at rest relative to one another under no forces other than those due to their own fields. The fundamental fact of the repulsion between two charges is thus lost sight of, and can only be recovered by reference to the structure of the electron after the manner of Lorentz and by the introduction of mechanical concepts such as that of the principle of least action†.

According to the Lorentz form of the theory an electrostatic field is impossible without the introduction of other than electric agencies, since otherwise there would be an uncompensated force on an electron contrary to the assumption stated above (§ 4). If there are no non-electromagnetic forces, then each electron will so move that the *total* force acting on it is zero, this force being due partly to the other electrons in the field, and partly due to the field of the moving electron itself.

6. It will be seen from this brief account therefore

(1) that in building up a theory from the differential equations of the field, it has not been found possible to remove entirely the mechanical conception of *force* or else *energy* as determining motion,

* *Aether and Matter*, pp. 94–8.

† Cf. *ibid.*

(2) that the electric and magnetic intensities are not quantities capable of direct observation, and

(3) that the field equations which have been stated are not completely determinative for a system of electrons without further assumptions.

These three remarks have an important bearing on the argument which follows.

7. On the Frame of Reference of the fundamental equations.

For the sake of clearness it is very necessary here to state what is assumed of the frame of reference or 'the aether' which is its objective equivalent. Naturally, in the theory as generally stated, we find traces of its history, and there is no doubt that the aether had been likened mentally, ever since the time of Faraday, to some kind of material medium.

M. Laue in his book on the *Principle of Relativity* remarks that it is a fundamental postulate of the electron theory, that "the aether is a *rigid body*, permeating everything, and so *defines* a unique system, to which the field equations are referred*." On the other hand Sir Joseph Larmor referring to the principle of relativity says† that if admitted in its universal form it implies "*a complete negation of any aethereal medium.*"

In view of these and similar statements, which indicate a general feeling of a need for an actually existing medium as that of which matter is but a partial manifestation, it is necessary to bear in mind that the conception of an objective aether filling all space has throughout its history been based entirely upon a mathematical analysis of the action *between matter and matter*, not upon any perceived properties of space apart from matter. The experiments which have been described above were undertaken with the sole purpose of giving a greater definiteness and concreteness to the conception of this medium by the discovery of its velocity relative to material objects.

* *Das Relativitätsprinzip*, p. 29.

† *Report of Mathematical Congress*, Cambridge, 1912, p. 214, footnote.

For we cannot, in our ordinary ways of thinking, speak of anything as *actually existing*, still less as being a *rigid body*, unless we can speak of it as having the geometrical properties of position and motion.

The remarkable fact is that all efforts have signally failed to establish these very properties. It is therefore far from justifiable, either to speak of the aether as a real *rigid* medium, or on the other hand to say that such a rigid medium is *required* as a frame of reference or to say that such a medium *defines* the frame of reference.

On the other hand the fact of the propagation in time of light and electromagnetic disturbances will always be a sufficient ground for a belief in some reality by means of which the transmission is effected. But all that is agreed as to this medium at present is that its state at any point, as far as electromagnetic propagation is concerned, may conveniently be represented by two vectors **e** and **h***. What is the relation of these two vectors to a possible distribution of velocity or strain throughout the medium it is for the moment irrelevant to discuss. It may well be remembered too in this connection that at the present moment the sufficiency of the existing conception of the aether to give an account of the results of modern experiments on radiation is being called in question.

Lorentz in expounding his theory says† "*We shall add the hypothesis that though the particles (electrons) may move, the aether always remains at rest.*" On examination, however, we see that the hypothesis that "the aether remains always at rest" does not enter into the discussion. The only thing that

* For further remarks on the concept of the aether see Larmor, *Aether and Matter*, p. 76, § 46, and especially the first paragraph of § 47, p. 78, from which we may quote the following passage: "...At the same time all that is known (or perhaps need be known) of the aether itself may be formulated as a scheme of differential equations defining the properties of a *continuum* in space, which it would be gratuitous to further explain by any complication of structure; though we can with great advantage employ our stock of ordinary dynamical concepts in describing the succession of different states thereby defined."

† *Theory of Electrons*, p. 11, ll. 18, 19.

suggests this is the fact that the field equations do not contain in the simple form in which they are stated any mention of velocity except that of the charges. Perhaps some support for the supposed necessity of the hypothesis lies in the similarity between the equations and those obtained by MacCullagh* in endeavouring to connect optical phenomena in a material media with the propagation of geometrical disturbances, in the medium, by elastic forces. The similarity only holds in case the elastic material medium which MacCullagh developed as the agency of transmission is considered to be as a whole at rest. The likening of the aether to a rigid medium at rest is however an example of the habitual practice, often helpful, sometimes misleading, of visualizing an abstraction in the form of an object of common experience†.

Just as in the case of Newton's Laws of Motion, then, *the definition of the frame of reference is in the laws which describe the phenomena, and not in a medium conceived to lie behind the phenomena.* We shall see in the next chapter that the definition is not unique, that the same degree of arbitrariness which exists in the dynamical case exists here, so that the electrical theory of matter brings us no nearer to a conception of an absolute frame of reference than does a purely mechanical theory.

* MacCullagh, *Collected Works*, p. 145.

† Compare the illustration developed by Larmor, *Aether and Matter*, Appendix E, of the properties of the aether, by a kinematic model in which the magnetic intensity is represented by the velocity of the medium and the electric intensity by the rotational displacement curl (ξ, η, ζ).

CHAPTER IV

CORRELATION OF STATIONARY AND MOVING SYSTEMS

1. We have now to consider an important property of the field equations of the electron theory, the discovery of which was entirely responsible for the formulation and discussion of the Principle of Relativity as it will be described in the subsequent chapters.

It would not serve much purpose here to describe in detail the way in which this property was discovered in the first place to a first approximation, then to a second approximation and finally in its completeness. References to this are given below *. A great stimulus to the work which led to the discovery was supplied by the null results of early experiments devised for the determination of the velocity of the earth relative to the aether, notably that of Mascart †, repeated later by Rayleigh. These were first order experiments, the Michelson-Morley experiment being the first attempt to investigate the possibility of a second order effect.

2. We assume first that 'a' frame of reference exists for which the fundamental equations hold.

In endeavouring to examine how electrical phenomena

* For the gradual evolving of this property of the equations see (1) Lorentz, *Versuch einer Theorie der elektrischen und optischen Erscheinungen in bewegten Körpern*, Leiden, 1895 to the first approximation; (2) Larmor, *Aether and Matter*, Cambridge, 1900, Chaps. X and XI to the second order of approximation; (3) Lorentz, *Amsterdam Proc.* 1904, p. 809; (4) Einstein, *Annalen der Physik*, 17 (1905). (1), (3) and (4) have been republished together, *Das Relativitätsprinzip*, Leipzig, 1913. See also Lorentz, *Theory of Electrons*, p. 198, footnote (1).

† Mascart, *Annales de l'école normale*, 3 (1874). Rayleigh, *Phil. Mag.* (6) 4 (1902), p. 215.

behave to an observer moving uniformly relatively to this frame of reference with a velocity v (which will be taken parallel to the axis of x), the first step will naturally be to take a new variable

$$x' = x - vt,$$

so that the observer has a constant coordinate x'.

When this is done the equations lose their original simplicity, but it is found that, by a further modification of the variables as follows, the original form can be recovered.

$$\left.\begin{array}{ll} \text{Putting} & X = \beta x' = \beta(x - vt), \\ \text{where} & \beta \equiv (1 - v^2/c^2)^{-\frac{1}{2}}, \\ \text{and also putting} & Y = y, \quad Z = z \\ \text{and} & T = \beta\left(t - \dfrac{vx}{c^2}\right) \end{array}\right\} \text{..................(A)},$$

we have, ψ being any function of (x, y, z, t),

$$\frac{\partial\psi}{\partial x} = \beta\left(\frac{\partial\psi}{\partial X} - \frac{v}{c^2}\frac{\partial\psi}{\partial T}\right), \qquad \frac{\partial\psi}{\partial t} = \beta\left(\frac{\partial\psi}{\partial T} - v\frac{\partial\psi}{\partial X}\right),$$

$$\frac{\partial\psi}{\partial y} = \frac{\partial\psi}{\partial Y}, \qquad \frac{\partial\psi}{\partial z} = \frac{\partial\psi}{\partial Z},$$

and the original equation (I) (p. 23) becomes in Cartesian coordinates

$$\frac{\beta}{c}\left(\frac{\partial e_x}{\partial T} - v\frac{\partial e_x}{\partial X}\right) + \frac{\rho u_x}{c} = \frac{\partial h_z}{\partial Y} - \frac{\partial h_y}{\partial Z},$$

$$\frac{\beta}{c}\left(\frac{\partial e_y}{\partial T} - v\frac{\partial e_y}{\partial X}\right) + \frac{\rho u_y}{c} = \frac{\partial h_x}{\partial Z} - \beta\left(\frac{\partial h_z}{\partial X} - \frac{v}{c^2}\frac{\partial h_z}{\partial T}\right),$$

$$\frac{\beta}{c}\left(\frac{\partial e_z}{\partial T} - v\frac{\partial e_z}{\partial X}\right) + \frac{\rho u_z}{c} = \beta\left(\frac{\partial h_y}{\partial X} - \frac{v}{c^2}\frac{\partial h_y}{\partial T}\right) - \frac{\partial h_x}{\partial Y}.$$

The second and third of these can be written

$$\frac{1}{c}\left(\frac{\partial E_y}{\partial T} + \rho u_y\right) = \frac{\partial H_x}{\partial Z} - \frac{\partial H_z}{\partial X} \text{..................}(\mathrm{I}_y'),$$

$$\frac{1}{c}\left(\frac{\partial E_z}{\partial T} + \rho u_z\right) = \frac{\partial H_y}{\partial X} - \frac{\partial X_x}{\partial Y} \text{..................}(\mathrm{I}_z'),$$

where

$$\left.\begin{aligned} E_x = e_x, \quad E_y = \beta\left(e_y - \frac{v}{c} h_z\right), \quad E_z = \beta\left(e_z + \frac{v}{c} h_y\right) \\ H_x = h_x, \quad H_y = \beta\left(h_y + \frac{v}{c} e_z\right), \quad H_z = \beta\left(h_z - \frac{v}{c} e_y\right) \end{aligned}\right\} \quad \ldots\ldots(\alpha).$$

These last equations, on solving, give

$$e_y = \beta\left(E_y + \frac{v}{c} H_z\right), \quad e_z = \beta\left(E_z - \frac{v}{c} H_y\right),$$

$$h_y = \beta\left(H_y - \frac{v}{c} E_z\right), \quad h_z = \beta\left(H_z + \frac{v}{c} E_y\right).$$

Making the substitutions (α) in the first equation of (I) we have

$$\frac{\beta}{c}\left\{\frac{\partial E_x}{\partial T} - v\left(\frac{\partial E_x}{\partial X} + \frac{\partial E_y}{\partial Y} + \frac{\partial E_z}{\partial Z}\right)\right\} + \frac{\rho u_x}{c} = \beta\left(\frac{\partial H_z}{\partial Y} - \frac{\partial H_y}{\partial Z}\right),$$

while (IV) becomes

$$\beta\left(\frac{\partial E_x}{\partial X} - \frac{v}{c^2}\frac{\partial E_x}{\partial T}\right) + \beta\left(\frac{\partial E_y}{\partial Y} + \frac{v}{c}\frac{\partial H_z}{\partial Y}\right) + \beta\left(\frac{\partial E_z}{\partial Z} - \frac{v}{c}\frac{\partial H_y}{\partial Z}\right) = \rho.$$

Multiplying this by v and adding to the preceding equation we have, remembering that $\beta^2 = (1 - v^2/c^2)^{-1}$,

$$\frac{1}{c}\left\{\frac{\partial E_x}{\partial T} + \beta\rho\,(u_x - v)\right\} = \frac{\partial H_z}{\partial Y} - \frac{\partial H_y}{\partial Z} \quad \ldots\ldots\ldots(\mathrm{I}_x'),$$

and also

$$\frac{\partial E_x}{\partial X} + \frac{\partial E_y}{\partial Y} + \frac{\partial E_z}{\partial Z} = \beta\rho\left(1 - \frac{v u_x}{c^2}\right) \quad \ldots\ldots\ldots(\mathrm{III}').$$

3. If we take the equations of Larmor in which $\rho = 0$ the equations (I′) and (III′) may be written in the form

$$\frac{1}{c}\frac{\partial \mathbf{E}}{\partial T} = \text{CURL}\,\mathbf{H},$$

$$0 = \text{DIV}\,\mathbf{E}^*,$$

which are of the same form as the original equations.

* $\text{CURL}\,\mathbf{H} = \left(\frac{\partial H_z}{\partial Y} - \frac{\partial H_y}{\partial Z}, \text{ etc.}\right)$, and so for $\text{DIV}\,\mathbf{E}$.

In the more general form if we put*

$$P = \beta\rho\left(1 - \frac{vu_x}{c^2}\right) \quad\text{.....................}(\beta),$$

$$\left.\begin{aligned} U_x &= \frac{u_x - v}{1 - \dfrac{vu_x}{c^2}} \\ U_y &= \frac{u_y}{\beta\left(1 - \dfrac{vu_x}{c^2}\right)} \\ U_z &= \frac{u_z}{\beta\left(1 - \dfrac{vu_x}{c^2}\right)} \end{aligned}\right\} \quad\text{.....................(B)},$$

the original form is also restored.

But it remains to attach a significance to the quantities so introduced in the equations (α), (β), (B).

On treating the equations

$$-\frac{1}{c}\frac{\partial \mathbf{h}}{\partial t} = \text{curl}\ \mathbf{e},$$

$$\text{div}\ \mathbf{h} = 0,$$

we find that exactly the same changes of variables lead to the equations

$$-\frac{1}{c}\frac{\partial \mathbf{H}}{\partial T} = \text{CURL}\ \mathbf{E} \quad\text{..............(II}'\text{)},$$

$$\text{DIV}\ \mathbf{H} = 0 \quad\text{.......................(IV}'\text{)}.$$

The change of variables contained in the equations (A), (α) above will henceforth be called the *Lorentz Transformation*, as its use in the present connection is chiefly due to his work.

* Lorentz, in his original paper (*Amst. Proc.* 1904), puts

$$P = \beta\rho;$$

$$U_x = u_x - v, \quad U_y = u_y, \quad U_z = u_z.$$

This does not make the correlation exact except for the electrostatic case, with which he deals more particularly, and further these values of U_x, U_y, U_z are inconsistent with those of X, Y, Z, T. See below, § 4, p. 54. Einstein was the first to give the complete transformation in this form.

4. The application of the Lorentz Transformation.

Taking first the equations of Larmor, assumed to be complete as determining the sequence of phenomena in a system, the transformation above described effects a correlation between two solutions of the same equations; that is, if the history of a certain system of electrons is given by the equations

$$\mathbf{e} = \mathbf{f}(x, y, z, t), \qquad \mathbf{h} = \mathbf{f}_1(x, y, z, t);$$

then on making the substitutions (A), (α) of the preceding section we obtain

$$\mathbf{E} = \mathbf{F}(X, Y, Z, T), \qquad \mathbf{H} = \mathbf{F}_1(X, Y, Z, T),$$

$\mathbf{F}$ and $\mathbf{F}_1$ being known in terms of $\mathbf{f}$ and $\mathbf{f}_1$ and the velocity v, and these new functions are solutions of the same field equations with the space-time variables (X, Y, Z, T); or reverting to the old variables

$$\mathbf{e}_1 = \mathbf{F}(x, y, z, t), \qquad \mathbf{h}_1 = \mathbf{F}_1(x, y, z, t)$$

give a new solution of the original equations.

If we compare the two solutions of the equations thus correlated we see

(i) That infinite values of $\mathbf{e}$ and $\mathbf{h}$ correspond to infinite values of $\mathbf{e}_1$ and $\mathbf{h}_1$, and vice versa.

It will be shewn later that if an infinity of $\mathbf{e}$ and $\mathbf{h}$ is of the kind arising from a point charge, then the corresponding infinity of $\mathbf{e}_1$ and $\mathbf{h}_1$ corresponds to an *exactly equal* point charge*.

(ii) That a point at rest in the original system corresponds to a point moving with velocity $-v$ in the derived system; for since

$$x = \beta(X + vT),$$

if x is constant $X + vT$ is constant.

* See p. 57. This is shewn by Larmor, *Aether and Matter*, p. 175, § 111, to be true to the order $(v/c)^2$ for the case where the charge in one of the systems is at rest, and is assumed to be true generally. This is as far as the treatment of the structure of an electron as a point allowed him to go as regards the general transformation. For the original treatment of the significance of this correlation see pp. 175 ff. of that work.

(iii) That the distance d between two stationary points on a line parallel to the axis of x is represented in the correlated system by a moving length $d(1 - v^2/c^2)^{\frac{1}{2}}$. For if x_1 and x_2 are independent of t we have, comparing values of X_1, X_2 for the same value of T,

$$x_1 = \beta(X_1 + vT),$$

$$x_2 = \beta(X_2 + vT),$$

giving $$x_2 - x_1 = \beta(X_2 - X_1).$$

The importance of (i) lies in the fact that in a theory of the constitution of matter in which the elementary constituent is a point charge, the magnitude of that ultimate charge is taken as an absolute and universal constant. The experimental evidence for this is mainly derived from Faraday's laws of electrolysis of solutions, except in so far as the approximate values determined from very different phenomena are in fairly close agreement with one another. But the hypothesis that the charge of an electron is a perfectly definite quantity is the only one that we have to replace the hypothesis that the mass of a material particle is a definite constant.

5. The significance of the correlation.

In the two systems thus correlated we have an exact correspondence of charges satisfying the same equations, and the suggestion is therefore made that *the two systems are equally capable of actually existing**.

If the first system were a system of charges at rest in the aether, the second would be a system of charges all moving with velocity v, the relative positions being unaltered save for a uniform contraction of the whole in the ratio $1 : (1 - v^2/c^2)^{\frac{1}{2}}$ in the direction parallel to the velocity.

If the first system were a material body at rest in the aether, assumed to be internally constituted of moving electrons whose

* Cf. Chapter XVI, p. 206, where the correlation of two systems is taken as a criterion of *equal probability*, and the bearing of this on the automatic contraction is further considered.

motions are determined by the field equations, the correlated system would *in its properties in bulk* be a body moving with velocity $-v$, its figure being that of the original body contracted in the same ratio.

From this it is argued that this contraction will actually and automatically take place if the body is actually set in motion from rest in the aether to a velocity $-v$, though of course it need hardly be said that the acceleration must be very gradual, or internal disturbances will be set up. The correspondence is between the normal state of the body at rest and the normal state of the body in motion. A reasonable explanation is thus given of the origin of the contraction suggested by FitzGerald and Lorentz to explain the null effect of the Michelson-Morley experiment.

6. The argument of Lorentz.

Historically Lorentz dealt with this question earlier than did Larmor, and his earlier treatment brings out a point which does not appear in the argument that has been just summarized.

We have seen that the fundamental equations of Lorentz have to be supplemented by some further assumption in order to make them complete, such as for example an assumption as to the shape of and the distribution of charge within the electron.

Now if an argument similar to the above is to be applied to the equations of Lorentz, the geometrical correlation must extend to these additional assumptions; an electron at rest must be correlated in all its properties with one in motion. The equations (B) must be a consequence of (A), and (β) must be consistent with the hypothesis of invariant charge*.

Thus Lorentz has to assume†, for example, that if the electron at rest is spherical, an electron in motion must be an oblate spheroid whose shorter axis is equal to the longer multiplied by $(1 - v^2/c^2)^{\frac{1}{2}}$.

* These statements are shewn to be justified below—see Chapter VI, § 4, pp. 55–7.

† This does not appear in the first discussion by Lorentz, it was not introduced until 1904 (see footnote, p. 31, (3)).

Further if other than electromagnetic forces are allowed as taking part in the determination of the configuration of a system, these must be assumed to be correlated for the two systems in the same manner as those of electromagnetic origin. The exact nature of this correlation is not elucidated by Lorentz*, in fact he does not pretend to give a complete theory of the matter. His treatment can only be said to point towards an explanation of the FitzGerald contraction. But quite enough emerges from his analysis to shew the important fact that, in the light of the known part which electrical forces play in the constitution of matter, we are *bound to recognize the possibility of changes in the shape and properties of material bodies when their velocity is altered*, and also that *the Newtonian conception of a rigid body as one having a permanent configuration independent of its velocity is one which is not even approximately realized unless that velocity is very small compared with that of light.*

7. The experiments of Rayleigh and Brace.

Following up this idea it occurred to Lord Rayleigh†, still thinking in terms of an actual fixed aether, that if an isotropic transparent body did actually experience a contraction as the result of its motion through the aether, it would seem to be no longer isotropic in its constitution, and that therefore there might result a double refraction if a beam of light traversed it in a direction obliquely to the direction of motion‡. On testing the matter he found no trace of such double refraction. Brace§ repeating the experiment with greater precision also found none; though the accuracy of his experiment was such that he would have detected an effect equal in magnitude to one

* See below, p. 159, § 8, and footnote.

† Rayleigh, *Phil. Mag.* (6), 4 (1902), p. 678. The expectation of a positive result in this or any other experiment was demurred to by Larmor when this paper was read at the British Association meeting at Belfast (Sept. 1902) on the ground of the argument up to second order phenomena given in *Aether and Matter*, Chap. XI.

‡ No such effect would of course be expected for a beam parallel to the direction of motion, as we should expect the optical properties to be symmetrical about this direction.

§ D. B. Brace, *Phil. Mag.* (6), 7 (1904), p. 317.

fiftieth part of that which would be produced by a contraction of the same magnitude produced by a mechanical pressure, or one hundredth part of that which Lorentz calculates would be produced by a simple closing up of all the particles of the refracting body without otherwise altering their properties*.

Thus if we are to admit the FitzGerald contraction as a means of explaining the Michelson-Morley experiment, then whatever the mechanism determining the optical properties of a transparent body, we must contemplate the possibility of this mechanism being subject to such compensating modifications as will neutralize the effect of the contraction. Lorentz, for example, is able by employing the conception of the electron mentioned in § 6, to explain the compensation as due to a difference in the effective mass of the electron for vibrations in different directions†.

8. The experiment of Trouton and Rankine‡.

With the same idea which prompted Lord Rayleigh's experiment, it occurred to F. T. Trouton that, taking a conducting body instead of an insulator, the FitzGerald contraction would produce an inequality of the conductivity of the body in different directions relative to the earth's motion. But experiments carried out by A. O. Rankine ‡ with the greatest of precision made it clear that there was no change in the current to the order of the fraction $(v/c)^2$ when the bar of metal conducting it was rotated in all possible directions.

Again therefore it follows that if the contraction exists, there are compensating modifications in whatever may be the mechanism of conduction.

9. The experiment of Trouton and Noble§.

This experiment is of a slightly different nature from those which have been described as it deals with an expected *mechanical* effect instead of an optical or electrical effect.

* Lorentz, *Theory of Electrons*, § 185.

† *Ibid.* p. 217. ‡ *Proc. Roy. Soc.* (A), 8 (1908), p. 428.

§ *Proc. Roy. Soc.* 72 (1903), p. 132. Also *Phil. Trans.* 202 (1903), p. 165.

The idea of the experiment is most simply expressed by taking the example of two equal and opposite charges, moving with the same velocity in parallel directions, the line joining them making an angle with the direction of the velocity which is not zero or a right angle.

Let the two charges be A and B and let them be moving with velocity v in the direction of the axis of x, taking A to be the origin and the coordinates of B to be $(\xi, \eta, 0)$, ξ and η being positive. Then the motion of A along the axis of x (this charge being supposed positive) produces a magnetic force at B parallel to the axis of z, and the negative charge moving across this field parallel to the axis of x is subject to a force parallel to the axis of y while the positive charge A is urged with an equal force in the opposite direction.

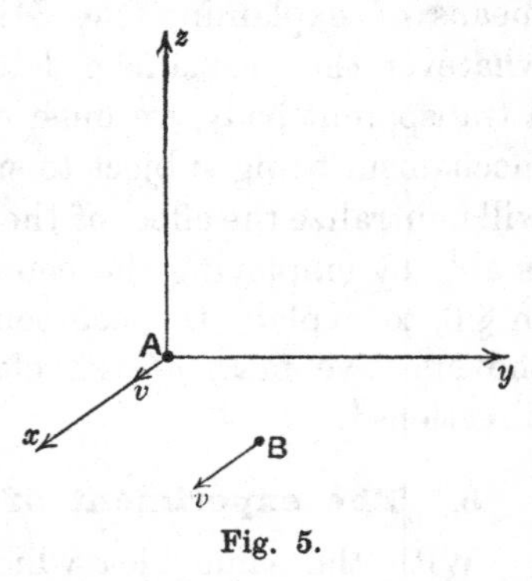

Fig. 5.

Thus if the charges were mounted on a bar which could turn about its centre, the bar would according to this rough theory be acted on by a couple tending to turn it into a direction at right angles to that of the motion of the charges, the couple being proportional to the square of either charge and to the square of the velocity.

In the actual experiment the charged bodies were the plates of a condenser, which was suspended by a thread, and a rotation of the condenser as a whole was expected when it was charged or discharged. But no rotation of the order expected was observed. The greatest deflection observed on the scale was ·36 cm., while the deflection indicated by the calculations was in some of the experiments as much as 6·8 cms.

In this experiment we are dealing with a direct comparison of forces indicated by existing electromagnetic theory with the ordinary mechanical property of the inertia of a rigid body. Thus the compensation observed in the region of optical and electrical effect appears to extend into that of mechanical forces. The way in which this compensation arises will be seen later when we come to deal with the mechanical bearing of the principle of relativity.

10. The significance of these experiments.

Thus it appears that if the hypothesis of the stagnant aether is to be maintained, we must admit that the correlation which we have seen to exist with reference to the fundamental electromagnetic equations must extend to at least some of that unexplored mechanism upon which the mechanical, optical and electrical properties of bodies depend.

It was with the hope of throwing further light on this fact that Lorentz developed his theory of the contracting electron in order to make more complete the correlation described above. But even then he had also to *assume* that the so-called 'elastic forces,' which in the theory of a dielectric body balance the electromagnetic forces acting on a particular electron and control its vibrations, were subject to a special modification to fit in with the general correlation. There are thus two distinct hypotheses *ad hoc*, neither of them following out of electromagnetic theory, each chosen for the sole reason of extending the correlation between different frames of reference which is possible in the case of the purely electromagnetic equations. That is, unless we are content with a general assumption that the electromagnetic equations in Larmor's form are sufficient to govern the whole of the electrical and optical properties concerned in the experiments referred to*, we have to admit that the general correlation between moving and stationary systems is more deep-seated than the electromagnetic equations themselves.

It is because the experimental evidence extends into regions where existing electrical theory is insufficient, that the principle of relativity becomes of importance as a supplementary and independent hypothesis.

* That is, to one part in 10^6 without analysing the physical structure of an electron.

CHAPTER V

EINSTEIN ON THE RELATIVITY OF SPACE AND TIME

1. We have now arrived at the stage at which Einstein introduced into the discussion an entirely novel point of view*, the point of view which is characteristic of the Principle of Relativity, a name which he was the first to introduce.

It will be well to recall the remarks made in the first chapter concerning the way in which the measurements of space and time are conditioned by the Newtonian Laws of Motion. From what was said there it will be seen, that if the Universe, as Laplace had once hoped†, is governed by those or some equivalent laws, as for example by the Principle of Least Action, the frame of reference which is necessary for the analytical expression of the laws is by no means unique. Within the realm of purely dynamical phenomena if the axes are such that (x, y, z, t) are valid space and time coordinates, then

$$x' = x - ut, \quad y' = y - vt, \quad z' = z - wt, \quad t' = t$$

give a set of space and time coordinates which are equally valid for the purpose of applying Newton's Laws of Motion or the Principle of Least Action in its ordinary dynamical form in which the kinetic energy of a particle is proportional to the square of its velocity. Thus if there are no phenomena of which we have cognizance which do not fall within the application of these laws, there is no meaning to be attached to the term *absolute position*, or *absolute velocity*. But certain quantities,

* *Annalen der Physik*, 17 (1905).

† "The discoveries of the human mind in mechanics and geometry, joined to that of universal gravitation, have brought it within reach of comprehending in the same analytical expressions the past and future states of the systems of the world." *Essai philosophique sur les probabilités*, p. 4, Paris, 1819.

force, acceleration, mass, which have the same value whatever set of coordinates is used, have from this point of view a concrete significance for the particular phenomena under observation.

Now we have seen in the last chapter, that there exists in the realm of electrodynamics a similar invariant property of the fundamental equations.

If we adopt Larmor's scheme in which those fundamental equations are the sole laws governing the changes in a system, it follows that we have absolutely no means of distinguishing between any set of space-time variables (x, y, z, t) *which is known to be valid, and the set* (x', y', z', t') *determined from them by the equations*

$$\left.\begin{aligned} x' = \beta(x - vt), \quad y' = y, \quad z' = z, \quad t' = \beta\left(t - \frac{vx}{c^2}\right) \\ \beta = (1 - v^2/c^2)^{-\frac{1}{2}}; \end{aligned}\right\} \ldots (A),$$

since, as has been pointed out, $\mathbf{e}$ and $\mathbf{h}$, the other variables in the equations, are not quantities capable of direct observation; they are only measured indirectly by observations of the motions of the electrons or their aggregates. They are in fact eliminated from the equations in obtaining any result capable of experimental verification. This point is so important, that it may be allowable to emphasize it further.

2. Coincidence the only exact form of observation.

The actual observations made are entirely dependent upon *simultaneous coincidences of position.* If *distances* are measured it is by securing a simultaneous coincidence of two points on a measuring scale with two points whose distance is required. This would include, for instance, the observation of interference fringes in optical experiments. *Forces* again cannot be measured directly, but only by means of an instrument depending on the same method. An *electric current* is measured by means of a galvanometer, the measurement being again made by observing a coincidence of a spot of light or a pointer with a mark on a scale. Every effort is made to eliminate any element of personal judgment as to distance or time interval, and only in so far as this is done are the results considered to be reliable.

The available test of the validity of the scheme of equations of the electromagnetic theory is therefore the comparison of *observed coincidences* with coincidences deduced from those equations by elimination of electric and magnetic intensities and, in the derived equations for material media, of such quantities as current strength, polarization, etc.; and when we speak of the *validity of the scheme of equations*, since they comprehend the whole of the phenomena which are under consideration, we include, not only the validity of the relations between the eliminated variables and the space and time variables, but also, what is far more important, *the validity of the system of measurement of space and time themselves**. Now in the transformations (A), to a given set of values of (x, y, z, t) corresponds a unique set of values (x', y', z', t'), so that what is described as a coincidence when the former quantities are employed as coordinates in space and time is equally well described as a coincidence when the latter are adopted in their place.

Every electromagnetic chain of phenomena is in a sense an electromagnetic clock and space-scale. It is in the mutual agreement of all such measuring instruments that the possibility of theoretical science lies. If it could be shewn that there were phenomena, not included within the scheme of the equation in question, which furnished an independent measure of space and time, the two measures would have to be consistent, or else one must be made fundamental, and the other treated as approximate, incorrect, or of only limited application.

To grant the completeness of Larmor's scheme is therefore to admit the impossibility of determining physically a unique system of variables (x, y, z, t) which may be called absolute space and time variables; or in other words it is to admit the impossibility of establishing the uniqueness of the aether—the term 'aether' being here understood merely as the frame of

* Cf. the discussion of the concept of time in Mach, *Science of Mechanics*, 2nd ed., 1902, pp. 222–6.

reference, to which the mind is apt to give a certain degree of substantiality. We must remember however that we have no guarantee that we can ignore the structure of the electron. The theory of Lorentz takes it into account, and equally well explains the results of experiments.

3. Discussion of the correlation of systems following Lorentz's equations.

Turning now to the scheme as proposed by Lorentz, which includes 'electric density' as an ultimate variable, it has been pointed out above that it is *not complete* as a purely electromagnetic scheme. If we therefore attempt to apply the same argument as that in the last section to a system constituted according to this scheme, we are bound, as was pointed out above, to assume that the correlation extends to all such additional hypotheses as are necessary to render the scheme a complete description of all the phenomena involved. If this is done we shall arrive at the same conclusion as that in the last section, that the frame of reference for the fundamental equations is not uniquely definable. On the other hand if the constitution of matter is determined partly by the equations of Lorentz, and partly by considerations which cannot be fitted into the correlation, the null effect of the experiments of Michelson and Morley, Rayleigh and Brace, Trouton, Noble and Rankine remain unexplained.

4. The Relativity of the measures of Space and Time.

The work of Lorentz and Larmor, in succeeding in rendering a partial account of what may be called *experimental relativity*, that is the failure to find evidence of absolute motion relative to the aether, has thus incidentally set on foot a revision of the concepts of space and time. Both these writers speak always of the aether as an objective reality, but assign to it properties which must for ever conceal the fact of uniform motion relative to it; although the idea of such a relative motion seems inseparable from an objective medium. Assuming the objective medium as defining a special frame of reference, in which the variable t is called the 'true time,' the correlated variable T for any other frame of reference is called by some such name as 'effective time.'

It is fundamental to the point of view associated with the principle of relativity that *if there is no means of physically determining which of all possible frames of reference is that which is at rest in the aether, then of all possible time variables no one is rightly named more than any other the 'true' time.*

This position was first clearly brought out by Einstein* in his paper of 1905, where, instead of beginning from a preconceived theory of the nature of electromagnetic phenomena and their place in the constitution of matter, he starts from a single hypothesis, namely that of the propagation of light with equal and constant velocity in all directions, not only relative to a single or unique frame of reference, but relative to any one of the triply infinite set of frames of reference which are equally valid for the purposes of dynamics.

This hypothesis was of course startling in view of the fundamental place that space and time had held in earlier thought. It was only possible in the light of such considerations as have been advanced above. Einstein saw that if the result of the Michelson-Morley experiment is accepted, our ordinary ideas as to the measures of time and space must be seriously modified.

5. The meaning of the term 'simultaneous.'

The first difficulty seems almost to have arisen straight out of the theory of that experiment.

Let A, B be two points supposed to be at a fixed distance apart, and let a ray of light be emitted from A and reflected at B so as to return to A.

Now, according to the customary view, if A, B are conceived to be at rest in the aether, if t_A and t_A' are the instants at which the signal starts from A and arrives back at A the

* Einstein, *Annalen der Physik*, 17 (1905), pp. 891–7.

moment of reflection t_B will be considered as instantaneous with the instant $(t_A + t_A')/2$.

We will write $$t_B = \tfrac{1}{2}(t_A + t_A').$$

On the other hand if A and B have a common velocity v in the direction AB and the corresponding times are denoted by τ_A, τ_B, τ_A' and the length AB by l, we have

$$\tau_B - \tau_A = \frac{l}{c - v},$$

$$\tau_A' - \tau_B = \frac{l}{c + v}.$$

Thus $$(\tau_B - \tau_A)(c - v) = (\tau_A' - \tau_B)(c + v),$$

giving $$\tau_B = \frac{\tau_A + \tau_A'}{2} + \frac{v(\tau_A - \tau_A')}{2c} \quad \ldots\ldots\ldots\ldots\ldots(1).$$

Thus in this case the time of reflection is not the arithmetic mean of the times of start and return.

We see therefore that *if the Michelson and Morley conclusion were universally valid the phrase* '**simultaneous occurrences at different points**' *would have no meaning until the velocity of those two points was stated*, for τ_B depends not only on τ_A and τ_A' but also on v.

6. Criticisms have been raised against this conclusion on the ground that we are not ultimately dependent on light signals for communication between different places, and that the ultimate measure of time in practice is the periodic rotation of the earth.

But of the latter objection it must be noted that the rotation of the earth relative to the fixed stars can only be measured or observed by the use of optical methods, and of the former it must be said that the objection to the suggested means of communication for the purpose of comparing time at two given places is one which would apply to any other means involving material media through which disturbances are conceived to be propagated with finite velocity.

If it could be shewn that for a stated velocity of the points A, B a standard of simultaneity could be set up by some physical means which differed from that expressed by the equation (1), the objection would hold; but the only other ways of doing this would involve communication by means of some material medium whose properties we have already learned to think of as capable of modification through space, so that again the criterion would probably depend on the velocity assigned. For the present therefore we seem compelled to fall back on Einstein's method of light signals for the control of the standards of time at different places.

There may possibly seem to be an alternative in gravitation as a connecting link between bodies independent of material media which is often said to be propagated with infinite velocity. But the laws of propagation of gravitation are entirely unknown, and in the circumstances, the only possibility seems to be to consider what the principle of relativity implies with regard to gravitational forces, and if possible to compare the implications with astronomical evidence*.

7. On the length of a moving body.

The definition of the 'length' of a moving rod as ordinarily understood would be the distance between two *stationary* points which are occupied *simultaneously* by the two ends.

If there is any doubt or ambiguity therefore about the terms 'stationary' and 'simultaneous,' a corresponding doubt is at once thrown on what is to be understood by the distance between two moving points.

If we assign a special value to the velocity v then the equation (1) becomes effective as a means of determining simultaneity and then the length of the rod becomes a definite quantity. But if v is not known, the length of the rod must be admitted either to be indefinite or else not to agree with the above definition which is taken for granted in all our usual modes of calculation.

* See Chapter XIII, pp. 178–180.

8. Transformations of space and time variables which leave the velocity of light unaltered.

From this point of view Einstein proceeds to consider what amount of arbitrariness in the space and time variables is consistent with the fundamental assumption of a given velocity of propagation of light in all directions.

Limiting the consideration to linear changes between the variables*, and to such changes that a point at rest in one system of variables corresponds to a point moving with uniform velocity in the other, a velocity which we will take to be v in the direction of the axis of x, we are limited to changes of the type

$$x' = k(x - vt), \quad y' = ly, \quad z' = lz,$$
$$t' = \alpha x + \beta y + \gamma z + \delta t.$$

As a consequence of these equations, the equation

$$x^2 + y^2 + z^2 = c^2t^2$$

is to lead to
$$x'^2 + y'^2 + z'^2 = c^2t'^2.$$

Substituting in the last equation we must have

$$k^2(x - vt)^2 + l^2y^2 + l^2z^2 = c^2(\alpha x + \beta y + \gamma z + \delta t)^2$$

equivalent to
$$x^2 + y^2 + z^2 = c^2t^2.$$

Comparing coefficients we see that

$$\beta = \gamma = 0,$$
$$k^2v + c^2\alpha\delta = 0,$$
$$k^2 - c^2\alpha^2 = l^2,$$
$$k^2v^2 - c^2\delta^2 = -c^2l^2,$$

which lead to
$$k = (1 - v^2/c^2)^{-\frac{1}{2}} l,$$
$$\delta = k, \quad \alpha = -\frac{kv}{c^2}.$$

* "On account of the homogeneous properties that we assign to space" is the reason he gives.

We shall for the present put $l = 1$ *, so that the equations of transformation are

$$x' = k(x - vt), \quad y' = y, \quad z' = z,$$

$$t' = k\left(t - \frac{vx}{c^2}\right),$$

k having the value $(1 - v^2/c^2)^{-\frac{1}{2}}$, and this is exactly the space-time transformation used by Lorentz and Larmor.

9. Einstein's conclusion then is that if light were the only means of communication between distant points by means of which a standard of simultaneity for events at different places could be set up, there would be exactly that arbitrariness in the measures of time and space that is suggested by the Lorentz transformation. In fact Lorentz' argument based on a hypothetical possibility as to the constitution of matter cannot be taken as more than a suggestion of what might be, whereas Einstein by limiting the consideration to a particular and less disputable class of phenomena reaches the important conclusion as to the relativity of the measures of space and time, and subsequently proceeds to the consideration of the bearing that this will have on the measures of electromagnetic magnitudes.

The correlation between two solutions of the fundamental equations described in the last chapter was there considered as connecting *two different systems* equally capable of existence in the same aether. To Einstein it simply develops the possibility of *two different descriptions of the same system* to two observers moving relative to one another with constant velocity v.

It may be said that the two arguments differ only in the point of view, but for the moment it is the new point of view that is vital. The older point of view was the outcome of the strivings of the mechanical school of physicists after an objective

* The effect of the factor l indicates only a uniform magnification of the scales of space and time, or what is the same thing, a change of units. It does not introduce any essential modification.

mechanical elastic aether and their adoption of a metaphysical conception of space and time. But when mechanics is placed in a derivative place and electrical theory in the fundamental position, when we are prepared to admit that many of the properties of matter are when analysed found to be compounded of electrical action, it becomes very necessary to consider with an open mind the possibility that some of what seem the most obvious modes of thought may require revision, and among other things to avoid the vicious circle of ideas involved in likening the aether to a species of matter with molar properties which are an idealization of the properties of matter as we imperfectly know them*.

It is felt by some physicists that to adopt the new point of view is to abolish the aethereal medium altogether, while putting nothing in its place—for apparently each observer may construct his own aether in which, for the moment, he is at rest. All objective reality thus seems to be taken away from it; and yet, paradoxically enough, the experimental results, indicating uniformly the absence of any unique aether, substantiate the whole electromagnetic theory, and indicate that the arguments of Lorentz and Larmor are based on a view of the constitution of matter which, in its main features, is true if incomplete. But it is only the same paradox as in the Newtonian dynamics. A definite frame of reference is there necessary for the analytical expression of the laws of motion. But when that expression has been formulated, the frame of reference is found to be far from unique.

Whether it will be possible in the future to adopt some other conception of the aether which is free from the objections attaching to its identification with the frame of reference cannot yet be foreseen. In the development of such a concept the new facts which are emerging in the theory of radiation will have to be taken into account. If it is to be consistent with the principle of relativity, a first requirement is that the velocity assigned

* See below, Chapter XV, Relativity and an Objective Aether.

to it shall be subject to the Einstein transformation of space and time when the frame of reference is altered.

It is indeed difficult to maintain a vivid conception of an aethereal medium in conjunction with the idea that we can never know what is the velocity of a point relative to it. But it is not so difficult to look upon time and space as mental modes under which we can catalogue our perceptions, and as being therefore not necessarily unique, but to a certain extent adaptable to the convenience of the moment. In the description of terrestrial phenomena, for example, the practical electrician invariably thinks of the earth as being at rest in the aether, and takes velocities relative to the earth as velocities relative to aether, just as much as the mechanics of every-day life is quite unconcerned with the enormous velocity of the earth in its orbit. It is only in astronomical work, both physical and descriptive, that the earth's velocity becomes of any account, and then another standard of rest is chosen.

CHAPTER VI

THE KINEMATICS OF EINSTEIN

1. It will be convenient now, adopting Einstein's point of view as to space and time, to develop some of the simpler consequences of it.

When the motion of bodies is described by means of relations between the coordinates (x, y, z) and the time t we shall say that the motion is in the frame S. If the relations are converted into relations connecting (x', y', z') with t', where

$$x' = \beta(x - vt), \quad y' = y, \quad z' = z, \quad t' = \beta\left(t - \frac{vx}{c^2}\right) \ldots \text{(A)},$$

we shall say that the motion is described in the frame S'.

2. The transformation is a reciprocal transformation, for on solving the equations (A) we obtain

$$x = \beta(x' + vt'), \quad y = y', \quad z = z', \quad t = \beta\left(t' + \frac{vx'}{c^2}\right) \ldots \text{(A')},$$

which is exactly of the same form with $-v$ substituted for v.

3. The FitzGerald contraction.

If two points P, Q are fixed in the frame S' their coordinates (x_1', y_1', z_1'), (x_2', y_2', z_2') are independent of t_1' and t_2'.

Hence applying (A)

$$x_1' = \beta(x_1 - vt_1),$$
$$x_2' = \beta(x_2 - vt_2).$$

If we wish to compare simultaneous positions in the frame S we put $t_1 = t_2$ obtaining thus

$$x_2' - x_1' = \beta(x_2 - x_1),$$

or

$$x_2 - x_1 = (x_2' - x_1')(1 - v^2/c^2)^{\frac{1}{2}},$$

together with

$$y_2 - y_1 = y_2' - y_1',$$
$$z_2 - z_1 = z_2' - z_1'.$$

Thus, by considering P, Q to be two points of a rigid body which is *at rest in* S', it follows that the length of any line in the body in the direction of the axis of x is estimated to be less in the frame S than the corresponding length in the frame S' in the ratio $1:(1 - v^2/c^2)^{\frac{1}{2}}$, the dimensions in directions at right angles to this being unaltered*.

4. The addition equations for velocities.

Consider next a moving point whose velocity in S is (u_x, u_y, u_z), and in S' is (u_x', u_y', u_z').

Then since for arbitrary changes δx, δy, δz, δt the corresponding changes in x', y', z', t' are given by

$$\left.\begin{aligned} &\delta x' = \beta(\delta x - v\delta t), \quad \delta y' = \delta y, \quad \delta z' = \delta z, \\ &\delta t' = \beta\left(\delta t - \frac{v\delta x}{c^2}\right), \end{aligned}\right\} \quad \ldots(\mathrm{A}_1),$$

we have

$$u_x' = \frac{\delta x'}{\delta t'} = \frac{\delta x - v\delta t}{\delta t - \dfrac{v\delta x}{c^2}} = \frac{u_x - v}{1 - \dfrac{vu_x}{c^2}},$$

and

$$u_y' = \frac{\delta y'}{\delta t'} = \frac{\delta y}{\beta\left(\delta t - \dfrac{v\delta x}{c^2}\right)} = \frac{u_y}{\beta\left(1 - \dfrac{vu_x}{c^2}\right)},$$

and similarly

$$u_z' = \frac{u_z}{\beta\left(1 - \dfrac{vu_x}{c^2}\right)}.$$

* For a discussion of the question as to whether the contraction would actually take place if the system were accelerated relative to a given frame of reference, see Chapter XVI, p. 206.

These are the equations for the transformation of velocities, and we see that they are identical with the substitutions (B) that it was found necessary to make in order to complete the correlation between two systems according to the equations of Lorentz (p. 34). The quantities (U_x, U_y, U_z) are now seen actually to be the velocities of a moving point as measured in the coordinates (X, Y, Z, T). This is a necessary part of the argument, a point which was not noted by Lorentz in 1904.

Taking the equation

$$\delta t' = \beta\left(\delta t - \frac{v\delta x}{c^2}\right),$$

and the corresponding one obtained from (A′)

$$\delta t = \beta\left(\delta t' + \frac{v\delta x'}{c^2}\right),$$

and multiplying, we obtain the result

$$1 = \beta^2\left(1 - \frac{vu_x}{c^2}\right)\left(1 + \frac{vu_x'}{c^2}\right),$$

or

$$\left(1 - \frac{vu_x}{c^2}\right)\left(1 + \frac{vu_x'}{c^2}\right) = 1 - \frac{v^2}{c^2} \quad \text{...........}(\mathrm{B}_1),$$

which is sometimes of use.

Further

$$\frac{1 - \dfrac{u'^2}{c^2}}{1 - \dfrac{u^2}{c^2}} = \frac{c^2\delta t'^2 - \delta x'^2 - \delta y'^2 - \delta z'^2}{c^2\delta t^2 - \delta x^2 - \delta y^2 - \delta z^2} \cdot \frac{\delta t^2}{\delta t'^2}$$

$$= \frac{\delta t^2}{\delta t'^2}$$

$$= \frac{1}{\beta^2\left(1 - \dfrac{vu_x}{c^2}\right)^2} \quad \text{........................}(\mathrm{B}_2).$$

5. Einstein's addition formula

$$u_x' = \frac{u_x - v}{1 - \dfrac{vu_x}{c^2}},$$

or what is the same thing

$$u_x = \frac{u_x' + v}{1 + \frac{vu_x'}{c^2}},$$

has some interesting and important implications.

(*a*) *If* $u_x' = 0$, *then* $u_x = v$, *and conversely if* $u_x = 0$, $u_x' = -v$.

Thus any point at rest in S' moves with velocity v in S; while any point at rest in S moves with velocity $-v$ in S'. The two frames of reference have equal and opposite velocities relative to each other.

(*b*) *Provided* $v < c$, u_x' is equal to, greater, or less than c according as u_x is equal to, greater, or less than c.

The facts contained in (*b*) are only a particular case of a more general theorem, namely, that the resultant velocity of a point in S' is equal to, greater, or less than c according as the resultant velocity in S is equal to, greater, or less than c. For we have identically from equations (A_1), p. 54,

$$\delta x'^2 + \delta y'^2 + \delta z'^2 - c^2\delta t'^2 = \delta x^2 + \delta y^2 + \delta z^2 - c^2\delta t^2.$$

The significance of this last equation will become clearer in the more general discussion to follow.

Hence if a point is moving with a velocity in S such that its displacement in a short time δt satisfies

$$\delta x^2 + \delta y^2 + \delta z^2 \lesseqgtr c^2\delta t^2,$$

then its displacement in the frame S' satisfies

$$\delta x'^2 + \delta y'^2 + \delta z'^2 \lesseqgtr c^2\delta t'^2.$$

(*c*) The addition equation may be written

$$\frac{c - u_x}{c - u_x'} = \frac{1 - v/c}{1 + vu_x'/c^2}.$$

Hence if u_x' and v are both positive and less than c,

$$\frac{c - u_x}{c - u_x'} < 1,$$

that is *the addition of a positive velocity* v *to the positive velocity* u_x' *gives a velocity nearer to* c *than* u_x. By the continual successive addition of velocities less than c in the same direction, the x component of the velocity would continually approach c but would never become equal to it.

Also considering the other components of velocity

$$\frac{u_y}{u_x} = \frac{u_y'}{\beta(u_x' + v)}.$$

If u_x' and v are both positive and less than c it follows that

$$\frac{u_y}{u_x} < \frac{u_y'}{u_x'},$$

that is not only does the component velocity u_x tend towards c by constant addition of other velocities, in a fixed direction, but the direction of the resultant velocity tends always towards the same direction.

6. Invariance of Charge.

We have seen above (p. 34) that when the charge is purely convected the transformation of the density necessary for invariance of the fundamental equations is

$$\mathrm{P} = \beta\rho(1 - vu_x/c^2) \quad \ldots\ldots\ldots\ldots\ldots\ldots(\beta),$$

and it was stated that this was consistent with *the invariance of electric charge*. This is shewn as follows.

If δa is a small element of volume moving with velocity (u_x, u_y, u_z) and δa_0 is the apparent volume to an observer moving with it

$$\delta a = \delta a_0 \left(1 - \frac{u^2}{c^2}\right)^{\frac{1}{2}}.$$

If δA is the volume to an observer to whom the velocity of the element is (U_x, U_y, U_z)

$$\delta A = \delta a_0 \left(1 - \frac{U^2}{c^2}\right)^{\frac{1}{2}}.$$

Thus
$$\frac{\delta A}{\delta a} = \frac{(1 - U^2/c^2)^{\frac{1}{2}}}{(1 - u^2/c^2)^{\frac{1}{2}}} = \frac{1}{\beta\left(1 - \dfrac{vu_x}{c^2}\right)},$$
by equation (B_2), p. 55.

Hence from (β)
$$\mathrm{P}\delta A = \rho \delta a,$$
that is, the charges in corresponding elements of volume are equal.

7. Failure of the ordinary concept of a rigid body.

So far we have referred only to points or to bodies which may be thought of as at rest in some particular frame of reference of the type allowed in this theory. But if we consider a body which has a rotation relative to any one frame, we see at once that certain difficulties are introduced into our ordinary conception of rigidity. In the Newtonian dynamics of a rigid body particles of the body which at any one instant lie on a straight line are always on a straight line, whatever the motion of the body, and whatever the frame of reference. This is not however consistent with the kinematics of a body as we are now viewing it.

A simple example will suffice to shew this. Consider a body which is thought of as rigid in the ordinary sense rotating uniformly about the axis of z in the frame S, that is let the coordinates of any point of it be given by
$$x = r\cos\omega t, \quad y = r\sin\omega t, \quad z = \text{const.}$$

Then in S' the coordinates are given by
$$\beta(x' + vt') = r\cos\beta\omega\left(t' + \frac{vx'}{c^2}\right),$$
$$y' = r\sin\beta\omega\left(t' + \frac{vx'}{c^2}\right).$$

Putting
$$x' + vt' = X, \quad y' = Y,$$
we have
$$\beta X = r\cos(\beta\omega X + \omega t'/\beta),$$
$$Y = r\sin(\beta\omega X + \omega t'/\beta).$$

The point then describes relative to the point $X = Y = 0$ an ellipse, but for different values of r we have

$$\frac{X}{Y} = \frac{1}{\beta} \cot \left(\beta \omega X + \frac{\omega t'}{\beta} \right),$$

which depends upon X as well as upon t' and is therefore not constant for any particular value of t', although x/y is constant for any particular value of t. Thus points which to one observer are on a straight line are not so to another.

It appears then that in the kinematical view which is here being presented, the ordinary conception of a rigid body requires a good deal of revision beyond the mere FitzGerald contraction*.

In order to prevent misconceptions it may be pointed out here that the FitzGerald contraction of a body is only suggested by the principle of relativity in the case where it is possible to choose one frame of reference in which the whole body is permanently at rest, and even then it has not been shewn that if the body is set in motion relative to that frame the contraction will automatically take place†. To apply it therefore, as some writers have endeavoured to do, to the small parts of a rigid body in rotation is to go beyond what is legitimate. It is not possible by means of the principle to say what will happen to a body in non-uniform motion in the way of deformation. But as we shall see later, and in fact as is shewn by the example just given, certain general assumptions about the motion of bodies are definitely inconsistent with the principle. Thus such suggestions for instance as that a plane disc, rotating about an axis at right angles to it, will buckle owing to the contraction of the faster moving outer part of the disc being greater than that of the part nearer to the centre are irrelevant and illogical. It is true that in order to conform with the principle of relativity

* A modified conception of rigidity consistent with the hypothesis of relativity has been developed by M. Born (*Ann. der Phys.* 30, and *Phys. Zeitschrift*, 11), but it is not possible to expound this in the present work. Reference must be made to the papers mentioned, and to other subsequent memoirs.

† See Chapter XVI, p. 206.

we may have to modify our specification of the stress in a rotating body. This will be indicated in a later chapter. But the FitzGerald contraction can only be expected of a body in uniform translatory motion and free from external stress.

8. The Doppler Effect.

If a plane wave of monochromatic light travelling along the axis of x is represented by

$$e_y = h_z = A \sin p\,(t - x/c),$$

and we apply to this the formulae (α), p. 33, we obtain

$$E_y = H_z = A\beta\left(1 - \frac{v}{c}\right)\sin \beta p\left\{t' + \frac{vx'}{c} - \frac{1}{c}(x' + vt')\right\}$$

$$= \beta A\left(1 - \frac{v}{c}\right)\sin \beta p\left(1 - \frac{v}{c}\right)\left\{t' - \frac{x'}{c}\right\},$$

which again gives a plane wave whose intensity is equal to that of the original wave multiplied by $\beta^2(1 - v/c)^2$, that is by $(c - v)/(c + v)$, and whose period is

$$\frac{2\pi}{p'} = \frac{2\pi}{\beta p\left(1 - \frac{v}{c}\right)} = \frac{2\pi}{p}\left(\frac{c+v}{c-v}\right)^{\frac{1}{2}}.$$

If we neglect $(v/c)^2$ this gives the ordinary Doppler effect of the change of period to an observer moving with velocity v in the direction of propagation of the light.

From this point of view the *true* period of the light, if such a term need be used at all, must be considered as the period observed by one who is at rest relative to the source of light.

9. Transverse Doppler Effect.

If we consider a beam of light travelling along the axis of y represented by

$$E_z = H_x = A \sin p\,(t - y/c),$$

the change of period reduces to one of the *second* order since

$$t - \frac{y}{c} = \beta\left(t' + \frac{vx'}{c^2} - \frac{y'}{\beta c}\right)$$

$$= \beta\left(t' - \frac{y' \cos\alpha - x' \sin\alpha}{c}\right),$$

where $\sin\alpha = v/c$.

Thus if τ is the period of the light as seen by one observer, and τ' the period as seen by a second observer who is seen by the first to have a velocity v perpendicular to the path of the light,

$$\tau' = \frac{\tau}{\beta} = \tau\,(1 - v^2/c^2)^{\frac{1}{2}}.$$

Thus there will be a second order change in the period.

It has been suggested* that a possibility of observing such an effect exists in the case of the light emitted by the canal rays. Experiments on the longitudinal Doppler effect shew that velocities up to $v/c = {\cdot}006$ exist in the particles constituting these rays. This would give a displacement of the spectral lines of the same order as the ordinary Doppler effect arising from a velocity of about 5×10^5 cms. per sec. (3 miles per sec.), that is for the sodium D-lines a shift of about ·1 Ångström unit. The experiment would however need to be carried out with great care in order to be quite sure that the ordinary longitudinal Doppler effect was not present owing to the direction of the rays being not quite perpendicular to the line of sight.

10. The Fresnel Convection-coefficient.

One of the most striking consequences of the Einstein kinematics is that it leads at once to the value suggested by Fresnel and experimentally verified by Fizeau for the velocity of a beam of light through a moving refracting medium.

* Laub, *Physical Review*, April, 1912; *Jahrbuch der Radioaktivität und Elektronik*, 7.

Suppose that to an observer moving with the medium the velocity of light is u and that the medium has a velocity v relative to a second observer.

Then, the velocity of the light relative to this second observer being u', the addition formula gives

$$u' = \frac{u+v}{1+uv/c^2}$$

$$= u + v\left(1 - \frac{u^2}{c^2}\right)$$

$$= u + v\left(1 - \frac{1}{n^2}\right),$$

neglecting $(v/c)^2$, n being the index of refraction of the light.

11. Modification of the convection-coefficient when the medium is dispersive.

If the medium is dispersive, n is a function of the period τ of the light, and this period will be different for the two observers. The value of n in the above formula is that corresponding to the period as seen by the observer moving with the medium, since it is relative to this observer that the velocity of the light in the medium is c/n.

If τ is the period of the light to this observer, and τ' to the other, the relation between τ' and τ is obtained as follows.

If ϕ represents any component of the light disturbance in the moving medium, we have

$$\phi = Ae^{\frac{2\pi i}{\tau}\left(t - \frac{x}{u}\right)},$$

x being a coordinate in the direction of the motion of the medium. But, using the relations between (x, t) and (x', t'),

$$x = \beta(x' - vt'), \quad t = \beta\left(t' - \frac{vx'}{c^2}\right),$$

we have
$$t-\frac{x}{u}=\beta\left\{t'\left(1+\frac{v}{u}\right)-x'\left(\frac{v}{c^2}+\frac{1}{u}\right)\right\}$$
$$=\beta\left(1+\frac{v}{u}\right)\left(t'-\frac{x'}{u'}\right).$$

Thus if we put
$$\tau'=\frac{\tau}{\beta\left(1+\frac{v}{u}\right)},$$
we have
$$\phi=Ae^{\frac{2\pi i}{\tau'}\left(t'-\frac{x'}{u'}\right)},$$
shewing that τ' is the period to the other observer.

We can now allow for the fact that the period of the light as it falls on the moving medium is τ and not τ'. Let n' be the index of refraction for light of period τ'. Then
$$n'=n+(\tau'-\tau)\frac{dn}{d\tau}$$
$$=n-\frac{v}{u}\tau\frac{dn}{d\tau}\quad\text{(neglecting } v^2\text{)}$$
$$=n\left(1-\frac{v\tau}{c}\frac{dn}{d\tau}\right).$$

Hence
$$u'=\frac{c}{n}+v\left(1-\frac{1}{n^2}\right)$$
$$=\frac{c}{n'}\left(1-\frac{v\tau}{c}\frac{dn}{d\tau}\right)+v\left(1-\frac{1}{n^2}\right)$$
$$=\frac{c}{n'}+v\left(1-\frac{1}{n'^2}\right)-\frac{v\tau}{n'}\frac{dn}{d\tau}.$$

The following figures shew that there is some little discrepancy between experiment and theory when the last term is taken into account.

Michelson and Morley* found for water with the sodium D-line that
$$u'-\frac{c}{n'}=\cdot 434v.$$

* *American Journal of Science*, 31 (1885).

The known figures for n' and $\frac{dn}{d\tau}$ give

$$v\left(1-\frac{1}{n'^2}\right) = \cdot 438v,$$

$$v\left\{\left(1-\frac{1}{n'^2}\right)-\frac{\tau}{n'}\frac{dn}{d\tau}\right\} = \cdot 451v.$$

Thus it appears that the theoretical term allowing for dispersion renders the agreement between theory and observation less satisfactory, though, since the possible error given in the account of the experiment was only $\pm \cdot 02v$, it cannot be said that the results are decisive. It is very desirable that the experiment should be repeated for the sake of testing this thoroughly.

In any case, however, the discrepancy is not connected with the change in scale of time in the ratio $1:\beta$ which is characteristic of the theory of relativity since it arises in the first order terms, whereas to the first order $\beta = 1$.

12. The Relativity of the electric and magnetic intensities.

Following up the same point of view which has been adopted in respect of space and time, Einstein concludes that in so far as the fundamental equations (I)—(IV) of Lorentz are valid as a set of relations descriptive of electromagnetic phenomena, to that extent it is impossible to say that $\mathbf{e}$, $\mathbf{h}$ rather than $\mathbf{E}$, $\mathbf{H}$ are the true electric and magnetic intensities, these quantities being connected by the transformation (α), p. 33. For all that is required of these intensities is that they should satisfy those equations; there is no further criterion.

Now it so happens that this point of view touches closely the experiments initiated by Kaufmann* in 1901 for the purpose of examining the mass of the small electrified particles moving with high velocities which are supposed to constitute the cathode rays, and carried further in 1905 in the case of the

* *Annalen der Physik*, 19 (1906).

similar particles moving with higher velocity still in the β-rays emitted by radium. It had been suggested* that the electrification of the particles would affect the apparent inertia of the particles in such a way that it would increase with the velocity, and this was what Kaufmann set out to examine.

In the arrangement of the experiment a stream of the particles, that is a narrow pencil of cathode or β-rays, was subjected to the action of electric and magnetic fields in a common direction perpendicular to the path of the rays, with the result that these rays were deflected from their straight path by measurable amounts. The deviation of the particles may be used to measure the intensity of the field, and the experiments thus have a bearing on Einstein's position as to the meaning of the intensity to an observer, in this case one of the particles, moving through the field. In order to make this clear it is necessary to consider the transformation of the acceleration of a moving point which is a consequence of the Lorentz Transformation.

13. Transformation of the acceleration of a moving point.

From the velocity transformation

$$u_x' = \frac{u_x - v}{1 - \frac{vu_x}{c^2}}, \qquad u_y' = \frac{u_y}{\beta\left(1 - \frac{vu_x}{c^2}\right)}, \qquad u_z' = \frac{u_z}{\beta\left(1 - \frac{vu_x}{c^2}\right)},$$

we obtain

$$\delta u_x' = \frac{\delta u_x\left(1 - \frac{v^2}{c^2}\right)}{\left(1 - \frac{vu_x}{c^2}\right)^2},$$

$$\delta u_y' = \frac{\delta u_y}{\beta\left(1 - \frac{vu_x}{c^2}\right)} + \frac{\delta u_x v u_y}{\beta c^2\left(1 - \frac{vu_x}{c^2}\right)^2}.$$

* The first suggestion of the electromagnetic inertia of a moving charged body was made by J. J. Thomson, *Phil. Mag.* 11 (1881), p. 229.

But, if δt, $\delta t'$ are corresponding elements of time, we have from the relation (B_2), p. 55,

$$\frac{\delta t'}{\delta t} = \frac{\left(1 - \frac{u^2}{c^2}\right)^{\frac{1}{2}}}{\left(1 - \frac{u'^2}{c^2}\right)^{\frac{1}{2}}}.$$

This quantity is of great importance in the transformation from S to S' and will be denoted by ϕ.

But also from the transformation (A) we obtain directly

$$\delta t' = \beta\left(\delta t - \frac{v}{c^2}\delta x\right),$$

so that if, following the point in its motion, we put $\delta x = u_x \delta t$, we have

$$\phi = \frac{\delta t'}{\delta t} = \beta\left(1 - \frac{vu_x}{c^2}\right).$$

Thus we may write

$$\frac{du_x'}{dt'} = \frac{1}{\beta^2\left(1 - \frac{vu_x}{c^2}\right)^2}\frac{du_x}{dt} \cdot \frac{dt}{dt'}$$

$$= \frac{1}{\phi^3}\frac{du_x}{dt},$$

and similarly
$$\frac{du_y'}{dt'} = \frac{1}{\phi^2}\frac{du_y}{dt} + \frac{\beta v u_y}{\phi^3 c^2}\frac{du_x}{dt},$$

$$\frac{du_z'}{dt'} = \frac{1}{\phi^2}\frac{du_z}{dt} + \frac{\beta v u_z}{\phi^3 c^2}\frac{du_x}{dt}.$$

Taking the simple case in which the point is momentarily at rest in S and putting (u_x, u_y, u_z) equal to zero, these reduce to

$$\frac{du_x'}{dt'} = \frac{1}{\beta^3}\frac{du_x}{dt},$$

$$\frac{du_y'}{dt'} = \frac{1}{\beta^2}\frac{du_y}{dt},$$

$$\frac{du_z'}{dt'} = \frac{1}{\beta^2}\frac{du_z}{dt}.$$

14. The Theory of Kaufmann's experiments*.

Suppose now that an electron with charge q is considered merely as a moving point having the properties of electric charge and inertia. Suppose that when placed at rest in a field in which the electric intensity is $\mathbf{e}$ it receives an acceleration given by

$$q\mathbf{e} = m\mathbf{f} \; \ldots\ldots\ldots\ldots \textit{Assumption (a)}.$$

For the present we do not ask why this should be so, or assume any theory as to the nature of the electron. This equation may be taken as a definition of $\mathbf{e}$.

Then as seen by an observer moving with velocity v along the axis of x relative to the electron, the electron is moving along the axis of x' with velocity $-v$ and its acceleration is now estimated as having components

$$\left.\begin{aligned} f_x' &= \frac{f_x}{\beta^3} = \frac{qe_x}{\beta^3 m} = \frac{qE_x}{\beta^3 m} \\ f_y' &= \frac{f_y}{\beta^2} = \frac{qe_y}{\beta^2 m} = \frac{q\left(E_y + \frac{v}{c} H_z\right)}{\beta m} \\ f_z' &= \frac{f_z}{\beta^2} = \frac{qe_z}{\beta^2 m} = \frac{q\left(E_z - \frac{v}{c} H_y\right)}{\beta m} \end{aligned}\right\} \ldots\ldots\ldots\ldots(1),$$

where $\mathbf{E}$, $\mathbf{H}$ are the electric and magnetic intensities in the system S'.

Assumption (*b*).

If now, following Lorentz, we call $q(\mathbf{E} + [\mathbf{uH}]/c)$ the *mechanical force* on the electron in the field, and if we make the second assumption that since $\mathbf{E}$, $\mathbf{H}$ satisfy the field equations in the frame S', they are the electric and magnetic intensities as measured by an observer in that frame, we may take the above equations as giving the accelerations produced by $\mathbf{E}$ and $\mathbf{H}$ in an electron having a velocity $-v$ along the axis of x.

* Clarendon type is used for vector quantities. For explanation of notation see pp. 83–4.

In the special arrangement used in the classic experiments of Kaufmann and his successors the directions of the electric and magnetic fields are the same and are perpendicular to the motion of a stream of electrons. Let the axis of y be taken in this direction. Thus we have from (1)

$$\left.\begin{aligned} qE_y &= \frac{mf_y'}{(1-v^2/c^2)^{\frac{1}{2}}} \\ \frac{qvH_y}{c} &= -\frac{mf_z'}{(1-v^2/c^2)^{\frac{1}{2}}} \end{aligned}\right\} \quad \text{.................. (2).}$$

In making the observations the stream of electrons impinges on a photographic plate at a known distance, say d, from the source. It is found that for given fields the trace which they leave on the plate is a curve starting out from the point where the stream impinges before the fields are excited. If (y, z) is a point on this curve, and t is the time of flight of an electron impinging at this point, we have, dropping the dashes for convenience,

$$y = \tfrac{1}{2} f_y t^2, \quad z = \tfrac{1}{2} f_z t^2, \quad d = vt.$$

Thus $$\frac{z}{y} = \frac{f_z}{f_y} = \frac{vH_y}{cE_z};$$

hence, given H_y and E_z, the velocity v can be determined for each point of the curve. This being done we have

$$\frac{q}{m} = \frac{f_y}{E_y(1-v^2/c^2)^{\frac{1}{2}}} = \frac{2y\,v^2}{d^2 E_y(1-v^2/c^2)^{\frac{1}{2}}} \quad \text{......... (3).}$$

The correctness of the assumptions (a) and (b) is now to be verified or otherwise by the constancy of the values of q/m calculated according to this equation. As the ratio v/c is found in some cases to reach the value ·9, this verification can be carried out with considerable accuracy, and on the whole the evidence is favourable*.

* Some figures are given later, Chap. XI, pp. 151–2. For a detailed discussion of results obtained by different experimenters under varying conditions, see Laub, *Jahrb. d. Rad. und Elektr.* 7 (1910), p. 405.

15. The equations (2) above may be written

$$\left.\begin{aligned} qE_x &= \frac{d}{dt}\left(\frac{mv_x}{\{1-(v_x^2+v_y^2+v_z^2)/c^2\}^{\frac{1}{2}}}\right) \\ q\left(E_y+\frac{v}{c}H_z\right) &= \frac{d}{dt}\left(\frac{mv_y}{\{1-(v_x^2+v_y^2+v_z^2)/c^2\}^{\frac{1}{2}}}\right) \\ q\left(E_z-\frac{v}{c}H_y\right) &= \frac{d}{dt}\left(\frac{mv_z}{\{1-(v_x^2+v_y^2+v_z^2)/c^2\}^{\frac{1}{2}}}\right) \end{aligned}\right\} \text{......(4)}$$

where after differentiation we put $v_y = v_z = 0$ and $v_x = -v$.

They lead to the equation

$$q(\mathbf{Ev}) = \frac{d}{dt}\left(\frac{mc^2}{\{1-(v_x^2+v_y^2+v_z^2)/c^2\}^{\frac{1}{2}}}\right) \text{.........(5)*.}$$

We have here a first suggestion of a modification required in the dynamics of a particle to be discussed in Chapter XI. The equation of momentum in the ordinary mechanical sense will remain satisfied if we assume for the 'momentum' the expression

$$\frac{m\mathbf{v}}{(1-\mathbf{v}^2/c^2)^{\frac{1}{2}}},$$

and the equation of energy will be satisfied if for the 'energy' we take

$$\frac{mc^2}{(1-\mathbf{v}^2/c^2)^{\frac{1}{2}}}.$$

But it will be observed that from the present point of view there is no need to use the mechanical terms 'force,' 'mass,' 'momentum' at all. The equations (1) (page 67) connect the acceleration with the electric and magnetic intensities, it being assumed (a) that for an electron momentarily at rest the acceleration is proportional to the electric intensity. Here m will naturally be called the 'mass' of the electron when at rest, but so far the mass of the moving electron remains undefined. The agreement of the equations (3) with experiment shews nothing

* See below, Chap. XIII, § 2, pp. 164–5.

about the dependence of *mass* on velocity until the definition of mass has been made; but it does confirm, as far as it goes, the theory that the equations on which the transformation of Einstein is based do in fact determine the motion of the electron, and that the quantities **E**, **H** are the electric and magnetic intensities as ordinarily interpreted. The foregoing discussion avoids by means of the primary assumptions (a), (b) all special theory as to the nature of the electron. If those assumptions are granted then the equations (3) follow no matter what be the speed of the electron. The experimental results therefore, in so far as they agree with the theory here given, supply a confirmation of those assumptions and of the kinematics on which the expressions for the acceleration of a moving point are based.

We shall defer the fuller discussion of the *mechanical interpretation* of the results to a subsequent chapter*. A table will there be given for reference shewing the closeness of the agreement of experiment with equations (3).

* Chapter XI, pp. 151–2.

CHAPTER VII

THE ELECTRON THEORY OF MATTER*

1. In the account that has been so far given, we have been dealing only with the equations of a hypothetical theory in which the only objects of consideration are the aether and free electricity. We must now pass on to the way in which this theory has been used to give an account of the electrical and magnetic properties of material bodies, and to the relation of these properties to the new point of view which has been outlined in the foregoing chapters.

The suggested conception of a material body is that it is pervaded, and partly or wholly constituted, by electric charges, which may be thought of, as has already been suggested, sometimes as point charges, sometimes as small nuclei of finite size. A molecule or atom is conceived to contain or to be composed of a group of such electrons, held together under their own mutual actions, and possibly in part by non-electromagnetic

* This chapter is inserted here partly as an illustration of the application of the kinematics of Einstein, partly as anticipating the results to be obtained later without reference to a constitutive theory of matter, and partly to supply readers with an account not otherwise easily accessible in English of the derivation of the electromagnetic equations for moving matter from the electronic conception. The treatment given is rather different in form from that usually given, and should be compared with that of Larmor, *Aether and Matter*, pp. 99 ff., and Lorentz, *Enzyk. der Math. Wiss.* V. 2, pp. 200–9. Vector notation is used, but as this is unfamiliar to many English students, explanatory notes are put at the foot of the pages, and reference may be made to the collected summary of results on pp. 83–4. The chapter may be omitted by non-mathematical readers.

forces. In addition to the electrons forming such groups there may (in the case of conducting bodies) be electrons moving between the molecules, or from one to another, so as to be capable of transporting a charge through any distance in the body. These two classes of electrons, 'free' and 'bound,' are treated separately.

The problem in hand is to explain the electromagnetic properties of matter in bulk in terms of the distribution and motion of these charges within it. For this purpose a process of averaging is undertaken, the most difficult part of which is the consideration of the average flux of electricity across an element of area. Owing to the fact that the electricity is supposed distributed in small nuclei, the actual density varies extremely rapidly from point to point within the body. But if the means of observation are only sufficiently refined to take account of elements of volume containing many electrons, these rapid changes will not be appreciable, and the distribution of electricity may without apparent alteration be replaced by another in which the changes are less rapid, provided that the total charge in such elements of volume is unaltered.

In the same way in dealing with the rate of flow of electricity across an element of area, the actual rate across an indefinitely small element varies extremely rapidly, but it is possible to imagine that the smoothed-out distribution of electricity gives the same flux across any element of area in any element of time, provided that many electrons cross the area in that time.

2. The free electrons.

Consider first then the *free electrons* which constitute the free charge on the body and which carry the current. We will suppose that we have replaced the actual distribution of charge by a continuous distribution giving the same effective density for elements of volume which are not too small, and the same effective flux for areas which are not too small.

Let the density of this distribution be ρ, the velocity of the matter be $\mathbf{u}$, and the velocity of the hypothetical distribution of charge relative to the matter be $\mathbf{w}$.

We have
$$\rho = \frac{\Sigma q}{\delta a}, \quad \rho\mathbf{w} = \frac{\Sigma q\mathbf{v}}{\delta a},$$
where δa is a small element of volume, and the summation extends over all the electron charges q within the volume, $\mathbf{v}$ being the velocity of an individual electron relative to the matter. Then the current density is represented by
$$\rho(\mathbf{u} + \mathbf{w}) = \rho\mathbf{u} + \rho\mathbf{w}.$$

The second term gives the average value of the stream of electricity across an element of area of the body which is small but sufficiently large for the purpose of smoothing out the rapid local variation in the actual flow. This term is called the *conduction current* and will be denoted by $\mathbf{j}$, while the first term is the *convection current* of free electricity.

3. The bound electrons.

Consider next those *bound electrons* which form a constituent part of the atoms of the body and which are therefore restricted in position to lie within a small region which moves with the body. Imagine those electrons which have negative charges to be in the same way replaced by a continuous distribution which gives a slow variation in density throughout the body, and which gives as before the same effective density and flux as that due to the electrons in question, and similarly with the positive electrons.

Assuming that when the body is free from electrical influences the total effective density at any point is zero, we may suppose that in this neutral state the body is permeated by two equal and opposite continuous distributions of densities ρ_0 and $-\rho_0$ arising from the positive and negative electrons respectively.

Here
$$\rho_0 = \frac{\Sigma_{(+)} q}{\delta a} \quad \text{and} \quad -\rho_0 = \frac{\Sigma_{(-)} q}{\delta a},$$
$\Sigma_{(+)}$ referring to the positive and $\Sigma_{(-)}$ to the negative electrons.

When by the action of external electrical forces the electrons are disturbed, the displacement of any one being represented by $\boldsymbol{\xi}$, let $\mathbf{r}_1$ be the displacement of the positive density ρ_0 relative to the body, so that

$$\mathbf{r}_1 = \frac{\Sigma_{(+)} q\boldsymbol{\xi}}{\Sigma_{(+)} q}.$$

Since we are considering *bound* electrons $\mathbf{r}_1$ may be taken to be small, but $\dot{\mathbf{r}}_1$ need not be small, since the charges might be in rapid rotary motion within the region to which they are confined.

Now if the displacement $\mathbf{r}_1$ varies from point to point in the displaced state the density ρ_0 at a point (x, y, z) of the body is changed to

$$\rho_0 - \operatorname{div}(\rho_0 \mathbf{r}_1),$$

since the displacement of charge outwards across a closed surface S' is

$$\iint \rho_0 (\mathbf{r}_1 d\mathbf{S}).$$

In the same way if the negative distribution $-\rho_0$ is shifted by $\mathbf{r}_2$ its density is altered to

$$-\rho_0 + \operatorname{div}(\rho_0 \mathbf{r}_2),$$

giving therefore together a density

$$\rho_1 = -\operatorname{div} \rho_0 (\mathbf{r}_1 - \mathbf{r}_2),$$

instead of zero density as in the neutral state.

If we introduce the vector $\mathbf{p}$ to stand for the product of ρ_0 and the relative shift $(\mathbf{r}_1 - \mathbf{r}_2)$, we have

$$\rho_1 = -\operatorname{div} \mathbf{p}.$$

4. The current due to the bound electrons.

The disturbed density of the positive electricity which represents the *bound positive* charges at the point (x, y, z) of space has been seen to be $\rho_0 - \operatorname{div} \mathbf{p}_1$, where

$$\mathbf{p}_1 = \rho_0 \mathbf{r}_1 = \frac{\Sigma_{(+)} q\boldsymbol{\xi}}{\delta a}.$$

To obtain the current due to the same distribution we need the velocity at the same point.

The velocity $\mathbf{u}+\dot{\mathbf{r}}$ (omitting the suffix 1 temporarily for convenience) gives the velocity of that element of charge which was when undisturbed at (x, y, z) but is now displaced by $\mathbf{r}$ from that point; here $\dot{\mathbf{r}}$ stands for the rate of increase of $\mathbf{r}$ for that particular element, that is

$$\dot{\mathbf{r}} = \frac{\partial \mathbf{r}}{\partial t} + (\mathbf{u}\nabla)\,\mathbf{r}^*.$$

The velocity at (x, y, z) is therefore

$$\begin{aligned}\mathbf{u}+\mathbf{w} &= (\mathbf{u}+\dot{\mathbf{r}}) - (\mathbf{r}\nabla)(\mathbf{u}+\dot{\mathbf{r}})\\ &= \mathbf{u} + \frac{\partial \mathbf{r}}{\partial t} + (\mathbf{u}\nabla)\,\mathbf{r} - (\mathbf{r}\nabla)\left(\mathbf{u} + \frac{\partial \mathbf{r}}{\partial t} + (\mathbf{u}\nabla)\,\mathbf{r}\right),\end{aligned}$$

in which the last term $(\mathbf{r}\nabla)(\mathbf{u}\nabla)\,\mathbf{r}$ may be neglected as being of the second order in the displacement $\mathbf{r}$ and its space rate of variation.

The current is given by the expression

$$(\rho_0 - \operatorname{div}\mathbf{p}_1)(\mathbf{u}+\mathbf{w}).$$

This expression may be analysed into the sum of the following parts:

(*a*) $$\rho_0\mathbf{u} \quad\text{.............................(i)}.$$

(*b*) $$\begin{aligned}\rho_0\frac{\partial \mathbf{r}_1}{\partial t} &= \frac{\partial \mathbf{p}_1}{\partial t} - \mathbf{r}_1\frac{\partial \rho_0}{\partial t}\\ &= \frac{\partial \mathbf{p}_1}{\partial t} + \mathbf{r}_1 \operatorname{div}(\rho_0\mathbf{u})\\ &= \frac{\partial \mathbf{p}_1}{\partial t} + \mathbf{p}_1\operatorname{div}\mathbf{u} + \mathbf{r}_1(\mathbf{u}\nabla)\,\rho_0\dagger,\end{aligned}$$

* $(\mathbf{u}\nabla)\,\mathbf{r}$ denotes the rate of change of the vector $\mathbf{r}$ as we move in the direction of $\mathbf{u}$ multiplied by the absolute value of $\mathbf{u}$; in Cartesian notation, if $\mathbf{r} = (\xi, \eta, \zeta)$, $(\mathbf{u}\nabla)\,\mathbf{r}$ is a vector whose components are

$$\left(u_x\frac{\partial \xi}{\partial x} + u_y\frac{\partial \xi}{\partial y} + u_z\frac{\partial \xi}{\partial z}\right), \text{ etc.}$$

† In Cartesian notation,

$$\begin{aligned}\operatorname{div}(\rho_0\mathbf{u}) &= \frac{\partial}{\partial x}(\rho_0 u_x) + \frac{\partial}{\partial y}(\rho_0 u_y) + \frac{\partial}{\partial z}(\rho_0 u_z)\\ &= \rho_0\left(\frac{\partial u_x}{\partial x} + \frac{\partial u_y}{\partial y} + \frac{\partial u_z}{\partial z}\right) + \left(u_x\frac{\partial \rho_0}{\partial x} + u_y\frac{\partial \rho_0}{\partial y} + u_z\frac{\partial \rho_0}{\partial z}\right)\\ &= \rho_0\operatorname{div}\mathbf{u} + (\mathbf{u}\nabla)\,\rho_0.\end{aligned}$$

See (1), p. 84.

remembering that, in the undisturbed state, the density is convected with the body, so that

$$\frac{\partial \rho_0}{\partial t} + \operatorname{div}(\rho_0 \mathbf{u}) = 0.$$

$$(c) \qquad -\mathbf{u} \operatorname{div} \mathbf{p}_1 + \rho_0 (\mathbf{u}\nabla) \mathbf{r} - \rho_0 (\mathbf{r}\nabla) \mathbf{u}$$
$$= -\mathbf{u} \operatorname{div} \mathbf{p}_1 - (\mathbf{p}_1\nabla) \mathbf{u} + \rho_0 (\mathbf{u}\nabla) \mathbf{r}_1.$$

Adding (*b*) and (*c*) together we obtain

$$(d) \qquad \frac{\partial \mathbf{p}_1}{\partial t} + \mathbf{p}_1 \operatorname{div} \mathbf{u} - \mathbf{u} \operatorname{div} \mathbf{p}_1 - (\mathbf{p}_1\nabla) \mathbf{u} + (\mathbf{u}\nabla) \mathbf{p}_1$$
$$= \frac{\partial \mathbf{p}_1}{\partial t} + \operatorname{curl} [\mathbf{p}_1 \mathbf{u}] \quad \text{..................(ii)*}.$$

Finally we have the terms involving $\frac{\partial \mathbf{r}_1}{\partial t}$:

$$(e) \qquad -\rho_0 (\mathbf{r}_1\nabla) \frac{\partial \mathbf{r}_1}{\partial t} - \operatorname{div} \mathbf{p}_1 \left\{ \frac{\partial \mathbf{r}_1}{\partial t} - (\mathbf{r}_1\nabla) \frac{\partial \mathbf{r}_1}{\partial t} \right\},$$

of which we neglect the third as of the second order leaving

$$-(\mathbf{p}_1\nabla) \frac{\partial \mathbf{r}_1}{\partial t} - \operatorname{div} \mathbf{p}_1 \,.\, \frac{\partial \mathbf{r}_1}{\partial t}.$$

Taking the x component of this expression, putting $\mathbf{r} = (\xi, \eta, \zeta)$, we have

$$-\frac{\partial \xi}{\partial t} \operatorname{div} \mathbf{p}_1 - (\mathbf{p}_1\nabla) \frac{\partial \xi}{\partial t}$$
$$= -\operatorname{div} \left(\frac{\partial \xi}{\partial t} \mathbf{p}_1 \right)$$
$$= -\operatorname{div} \left\{ \rho_0 \left(\xi \frac{\partial \xi}{\partial t}, \ \eta \frac{\partial \xi}{\partial t}, \ \zeta \frac{\partial \xi}{\partial t} \right) \right\}$$
$$= -\operatorname{div} \left[\frac{1}{2} \frac{\partial}{\partial t} \{\rho_0 \xi^2, \ \rho_0 \xi\eta, \ \rho_0 \xi\zeta\} + \frac{1}{2} \rho_0 \left\{ 0, \ \eta \frac{\partial \xi}{\partial t} - \xi \frac{\partial \eta}{\partial t}, \ \zeta \frac{\partial \xi}{\partial t} - \xi \frac{\partial \zeta}{\partial t} \right\} \right],$$

* Since

$$\frac{\partial}{\partial y} (p_x u_y - p_y u_x) - \frac{\partial}{\partial z} (p_z u_x - p_x u_z)$$
$$= p_x \left(\frac{\partial u_x}{\partial x} + \frac{\partial u_y}{\partial y} + \frac{\partial u_z}{\partial z} \right) - u_x \left(\frac{\partial p_x}{\partial x} + \frac{\partial p_y}{\partial y} + \frac{\partial p_z}{\partial z} \right)$$
$$+ \left(u_x \frac{\partial p_x}{\partial x} + u_y \frac{\partial p_x}{\partial y} + u_z \frac{\partial p_x}{\partial z} \right) - \left(p_x \frac{\partial u_x}{\partial x} + p_y \frac{\partial u_x}{\partial y} + p_z \frac{\partial u_x}{\partial z} \right).$$

See (2), p. 84.

in which $\frac{\partial \rho_0}{\partial t}$ is neglected since it leads only to terms of higher order.

The assumption of Lorentz is now that $\rho_0 \xi^2$, $\rho_0 \xi\eta$, $\rho_0 \xi\zeta$, etc., remain effectively constant, the rapid motion of the electrons within an atom being orbital. If therefore we neglect the first term in the last expression we are left with

$$-\operatorname{curl} \tfrac{1}{2}\rho_0 \left[\mathbf{r}_1 \frac{\partial \mathbf{r}_1}{\partial t}\right]^*,$$

which, neglecting again quantities of the second order, can be written

$$-\operatorname{curl} \tfrac{1}{2}\rho_0 [\mathbf{r}_1 \dot{\mathbf{r}}_1] \text{ ....(iii).}$$

Adding all the expressions (i), (ii), (iii) so obtained we may write therefore for the total flux due to the positive *bound* electricity

$$\rho_0 \mathbf{u} + \frac{\partial \mathbf{p}_1}{\partial t} + \operatorname{curl} [\mathbf{p}_1 \mathbf{u}] - c \operatorname{curl} \mathbf{m}_1,$$

where $$c\mathbf{m}_1 = \tfrac{1}{2}\rho_0 [\mathbf{r}_1 \dot{\mathbf{r}}_1].$$

In the same way for the negative electricity we should obtain the expression

$$-\rho_0 \mathbf{u} + \frac{\partial \mathbf{p}_2}{\partial t} + \operatorname{curl} [\mathbf{p}_2 \mathbf{u}] - c \operatorname{curl} \mathbf{m}_2,$$

where $\mathbf{p}_2 = -\rho_0 \mathbf{r}_2$, and $\mathbf{m}_2 = -\tfrac{1}{2}\rho_0 [\mathbf{r}_2 \dot{\mathbf{r}}_2]$.

Adding these together we get for the total current

$$\frac{\partial \mathbf{p}}{\partial t} + \operatorname{curl} [\mathbf{p}_1 \mathbf{u}] - \operatorname{curl} \mathbf{m},$$

where as before $\mathbf{p} = \mathbf{p}_1 + \mathbf{p}_2$ and also $\mathbf{m} = \mathbf{m}_1 + \mathbf{m}_2$†.

The quantity $\mathbf{p}$ is called the 'polarization,' and $\mathbf{m}$ the 'magnetization.'

* Since the x component of curl $[\mathbf{ab}]$ is equal to

$$\frac{\partial}{\partial y}(a_x b_y - a_y b_x) - \frac{\partial}{\partial z}(a_z b_x - a_x b_z) = \operatorname{div}\{0,\ (a_x b_y - a_y b_x),\ -(a_z b_x - a_x b_z)\}.$$

This is a particular case of the theorem (4), p. 84.

† It is perhaps not quite clear at first sight what is the relation of the quantity $\mathbf{m}$ defined for the smoothed-out distribution to the motion of the

5. We can now turn back to the fundamental equations of Lorentz, namely

$$\frac{1}{c}\left(\frac{\partial \mathbf{e}}{\partial t} + \mathbf{i}\right) = \text{curl } \mathbf{h},$$

$$-\frac{1}{c}\frac{\partial \mathbf{h}}{\partial t} = \text{curl } \mathbf{e},$$

$$\text{div } \mathbf{e} = \rho,$$

$$\text{div } \mathbf{h} = 0,$$

where $\mathbf{i} = \rho(\mathbf{u} + \mathbf{w})$, $\mathbf{u} + \mathbf{w}$ being now the velocity of the moving charge, instead of $\mathbf{u}$ as in the original notation. We have obtained above an expression for the average value of $\mathbf{i}$ through a small volume round any point.

The processes of space averaging through a *fixed* volume in space and differentiating with regard to the time are clearly commutative so that the above equations may be taken to hold for the averaged distribution. The average value of $\mathbf{h}$ will be denoted by $\mathbf{b}$, while we retain the symbol $\mathbf{e}$ after averaging.

Substituting the expression for the average value of $\mathbf{i}$ as found above we have

original nuclei. It is clear that the averaged distribution is not uniquely determined by the conditions that the effective density and flux shall be the same for fairly small elements as in the actual distribution—that in fact the substitution of an averaged distribution may be made in many ways. The analysis in the text holds for all these, so that on the assumption of the same average flux for all distributions, the average curl of the quantity $\mathbf{m}$ must be the same for all.

If we suppose that the original distribution is one in which the density though fluctuating rapidly is everywhere finite the original distribution is one of the many satisfying the required condition. Thus we are bound to assume that the average curl $\mathbf{m}$ of the smoothed-out distribution is the same as that of the actual. This will certainly be satisfied if the $\mathbf{m}$ of the smoothed-out distribution is the averaged $\mathbf{m}$ of the actual, and we may make this assumption of the averaged distribution in addition to those already made without fear of inconsistency. That is, we may put

$$\mathbf{m} = \frac{1}{\delta V}\iiint \rho[\mathbf{r}\dot{\mathbf{r}}]/2c \,.\, dV \text{ or } = \frac{1}{\delta V}\Sigma\frac{q[\mathbf{r}\dot{\mathbf{r}}]}{2c},$$

where δV is the small volume through which we average.

$$\frac{1}{c}\left(\frac{\partial \mathbf{e}}{\partial t} + \rho\mathbf{u} + \mathbf{j} + \frac{\partial \mathbf{p}}{\partial t} + \text{curl}\,[\mathbf{pu}] + c\,\text{curl}\,\mathbf{m}\right) = \text{curl}\,\mathbf{b},$$

$$-\frac{1}{c}\frac{\partial \mathbf{b}}{\partial t} = \text{curl}\,\mathbf{e},$$

$$\text{div}\,\mathbf{e} = \rho - \text{div}\,\mathbf{p},$$

$$\text{div}\,\mathbf{b} = 0.$$

Writing $$\mathbf{d} = \mathbf{e} + \mathbf{p}$$

and $$\mathbf{h} = \mathbf{b} - \mathbf{m}$$

the equations become

$$\frac{1}{c}\frac{\partial \mathbf{d}}{\partial t} + \rho\mathbf{u} + \mathbf{j} + \text{curl}\,[\mathbf{pu}] = \text{curl}\,\mathbf{h} \quad\text{............(I)},$$

$$-\frac{1}{c}\frac{\partial \mathbf{b}}{\partial t} = \text{curl}\,\mathbf{e} \quad\text{....................(II)},$$

$$\text{div}\,\mathbf{d} = \rho \quad\text{......................(III)},$$

$$\text{div}\,\mathbf{b} = 0 \quad\text{......................(IV)},$$

and this is the set of equations as proposed by Lorentz for moving bodies.

6. The application of the Einstein kinematics to the electron theory of matter.

The quantities ρ, $\mathbf{j}$, $\mathbf{p}$ and $\mathbf{m}$ in the foregoing discussion depend only upon the distribution and motion of the charge nuclei. Now, as will be shewn immediately, in the Lorentz transformation the magnitudes of the charges are unaltered, while their positions and motions are subject to certain purely geometrical transformations. If the corresponding quantities P, $\mathbf{J}$, $\mathbf{P}$, $\mathbf{M}$ are then formed from the resulting distribution, we shall find that they are expressible in terms of ρ, $\mathbf{j}$, $\mathbf{p}$, $\mathbf{m}$ and the relative velocities of the two systems of reference.

The averaged quantities $\mathbf{e}$, $\mathbf{b}$ will clearly be subject to the same transformation (α) as $\mathbf{e}$, $\mathbf{h}$ before averaging, and with the set of transformations for ρ, $\mathbf{j}$, $\mathbf{p}$, $\mathbf{m}$ we shall have a complete set of transformations which must leave the final equations (I, II, III, IV) for the moving matter invariant. These transformations for ρ, $\mathbf{j}$, $\mathbf{p}$, $\mathbf{m}$ will now be obtained.

(*a*) *Transformation of density and current.*

We have seen above that the condition of conservation of charge is

$$\frac{\mathrm{P}}{\rho} = \frac{\delta a}{\delta A} = \beta\left(1 - \frac{v u_x}{c^2}\right),$$

u_x being the component of the velocity with which the charge ρ is moving.

If however the charge is not convected as a whole but is carried by a multitude of moving electrons with different velocities, this equation will only apply to a particular group of charges moving all with the same velocity $\mathbf{u}$.

Suppose now that we are considering a material body moving with velocity $\mathbf{u}$ and that we consider for a moment a group of electrons moving with velocity $\mathbf{u} + \mathbf{w}$.

Then the transformation for the average density arising from this group will be

$$\mathrm{P} = \beta\rho\left\{1 - \frac{v(u_x + w_x)}{c^2}\right\}$$

$$= \beta\rho\left(1 - \frac{v u_x}{c^2}\right) - \beta\frac{v}{c^2}\rho w_x,$$

and here ρw_x is that part of the *conduction* current which is due to this group of electrons.

Thus on summing for all groups of all velocities we obtain the equation

$$\mathrm{P} = \beta\rho\left(1 - \frac{v u_x}{c^2}\right) - \beta\frac{v}{c^2} j_x \quad \ldots\ldots\ldots\ldots\ldots (\beta_1),$$

where j_x is the total conduction current due to electrons of all velocities.

Consider similarly the conduction current due to a group of common velocity

$$\mathbf{J} = \frac{\Sigma q \mathbf{W}}{\delta A}$$

and $$\mathbf{j} = \frac{\Sigma q \mathbf{w}}{\delta a}.$$

Here $\mathbf{W}$ and $\mathbf{w}$ are velocities relative to the moving matter.

The application of the addition equation to the *relative* velocity of two moving points gives

$$W_x = w_x \Big/ \beta^2 \left(1 - \frac{vu_x}{c^2}\right)\left(1 - \frac{v(u_x + w_x)}{c^2}\right),$$

$$W_y = w_y \Big/ \beta \left(1 - \frac{vu_x}{c^2}\right) - vu_y w_x \Big/ \beta c^2 \left(1 - \frac{vu_x}{c^2}\right)\left(1 - \frac{v(u_x + w_x)}{c^2}\right).$$

Remembering that

$$\frac{\delta a}{\delta A} = \beta \left(1 - \frac{v(u_x + w_x)}{c^2}\right),$$

this gives at once

$$J_x = \frac{\Sigma q w_x}{\beta \left(1 - \frac{vu_x}{c^2}\right) \delta a}$$

$$\left.\begin{aligned} &= \frac{j_x}{\beta\left(1 - \frac{vu_x}{c^2}\right)}, \\ \text{and}\quad J_y &= \frac{\Sigma q w_y}{\delta a} - \frac{\Sigma q v u_y u_x}{c^2\left(1 - \frac{vu_x}{c^2}\right)\delta a} \\ &= j_y - \frac{vu_y}{c^2\left(1 - \frac{vu_x}{c^2}\right)} j_x, \\ \text{and similarly}\quad J_z &= j_z - \frac{vu_z}{c^2\left(1 - \frac{vu_x}{c^2}\right)} j_y \end{aligned}\right\} \quad \ldots\ldots\ldots\ldots (\gamma);$$

and, since the velocity $\mathbf{w}$ does not occur in these equations, they hold for all velocities of the electrons, and therefore for the total current.

(*b*) *The Polarization and Magnetization.*

In a similar way, though the analysis is necessarily longer*, from the definitions

$$\mathbf{p} = \frac{\Sigma q\mathbf{r}}{\delta a}, \quad \mathbf{m} = \frac{\Sigma q\,[\mathbf{rw}]}{2c\,\delta a}$$

we obtain transformations which, using the suffix 0 for the values at a point at which the material velocity $\mathbf{u}$ is taken to be zero†, are

$$\begin{aligned} P_x &= (p_x)_0, & M_x &= (m_x)_0, \\ P_y &= \beta\,(p_y + vm_z/c)_0, & M_y &= (m_y/\beta)_0, \\ P_z &= \beta\,(p_z - vm_y/c)_0, & M_z &= (m_z/\beta)_0. \end{aligned}$$

The above transformations, together with

$$\begin{aligned} E_x &= (e_x)_0, & B_x &= (b_x)_0, \\ E_y &= \beta\,(e_y - vb_z/c)_0, & B_y &= \beta\,(b_y + ve_z/c)_0, \\ E_z &= \beta\,(e_z + vb_y/c)_0, & B_z &= \beta\,(b_z - ve_y/c)_0, \end{aligned}$$

must leave the equations (I—IV), p. 79, unaltered‡.

One or two consequences of these transformations may be noted.

Putting $\mathbf{u} = 0$ in (β_1), p. 80, and (γ), p. 81, we have

$$\mathrm{P} = \beta\left\{\rho_0 - \frac{v}{c^2}(j_x)_0\right\},$$

$$\mathbf{J} = (j_x/\beta,\ j_y,\ j_z)_0,$$

and here $-v$ is the velocity of the medium to the observer of P and $\mathbf{J}$.

If δA, δa_0 are corresponding elements of volume of the matter

$$\delta A = \delta a_0/\beta.$$

* See *Lond. Math. Soc. Proceedings*, Feb. 9, 1911, pp. 123–6, for the proofs of the statements.

† For convenience the form of the transformation is only given for the case where in one system the point in question is at rest. The general transformation could equally well be obtained directly or in two steps through an intermediate comparison with a system in which the point is at rest.

‡ Cf. below, p. 115.

Thus $$\mathrm{P}\delta A = \rho_0 \delta a_0 - \frac{v}{c^2}(j_x)_0\, \delta a_0.$$

That is, the *apparent charges* $\mathrm{P}\delta A$, $\rho_0 \delta a_0$ *in corresponding elements of volume of the moving matter are not equal unless the conduction current vanishes**. This is not a contradiction of the fundamental result that the charge of any given element of electricity is an invariant, but rather, as has been shewn, a consequence of that result.

APPENDIX

Summary of notation and of some results in Vector Analysis, and their equivalents in Cartesian Notation.

$$\mathbf{a} = (a_x, a_y, a_z), \quad \mathbf{b} = (b_x, b_y, b_z).$$

Absolute value. $$|\mathbf{a}| = (a_x^2 + a_y^2 + a_z^2)^{\frac{1}{2}}.$$

Scalar product. $$(\mathbf{ab}) = a_x b_x + a_y b_y + a_z b_z = |\mathbf{a}|\,|\mathbf{b}| \cos\theta,$$

where θ is the angle between **a** and **b**.

Vector product. $$[\mathbf{ab}] = (a_y b_z - a_z b_y,\ a_z b_x - a_x b_z,\ a_x b_y - a_y b_x),$$
$$|[\mathbf{ab}]| = |\mathbf{a}|\,|\mathbf{b}| \sin\theta.$$

$[\mathbf{ab}]$ is perpendicular to **a** and to **b**, and the directions of **a**, **b**, $[\mathbf{ab}]$ are in right-handed screw order.

Divergence.

$$\operatorname{div} \mathbf{a} = \frac{\partial a_x}{\partial x} + \frac{\partial a_y}{\partial y} + \frac{\partial a_z}{\partial z} = \operatorname*{Lt}_{V=0} \frac{1}{V}\iint (\mathbf{a}\, d\mathbf{S}),$$

where the double integral is taken over the surface enclosing V.

Rotation.

$$\operatorname{curl} \mathbf{a} = \left(\frac{\partial a_z}{\partial y} - \frac{\partial a_y}{\partial z},\ \frac{\partial a_x}{\partial z} - \frac{\partial a_z}{\partial x},\ \frac{\partial a_y}{\partial x} - \frac{\partial a_x}{\partial y}\right) = [\nabla \mathbf{a}],$$

where ∇ is the operator $\left(\frac{\partial}{\partial x}, \frac{\partial}{\partial y}, \frac{\partial}{\partial z}\right)$.

* Cf. pp. 57–8.

The operator $(\mathbf{a}\Delta)$.

$$(\mathbf{a}\Delta)\,\phi = a_x\frac{\partial\phi}{\partial x} + a_y\frac{\partial\phi}{\partial y} + a_z\frac{\partial\phi}{\partial z}.$$

$$(\mathbf{a}\Delta)\,\mathbf{b} = \begin{bmatrix} a_x\frac{\partial b_x}{\partial x} + a_y\frac{\partial b_x}{\partial y} + a_z\frac{\partial b_x}{\partial z} \\ a_x\frac{\partial b_y}{\partial x} + a_y\frac{\partial b_y}{\partial y} + a_z\frac{\partial b_y}{\partial z} \\ a_x\frac{\partial b_z}{\partial x} + a_y\frac{\partial b_z}{\partial y} + a_z\frac{\partial b_z}{\partial z} \end{bmatrix}.$$

Theorems.

(1) $\operatorname{div}(\phi\mathbf{a}) = \phi\operatorname{div}\mathbf{a} + (\mathbf{a}\nabla)\,\phi.$

$$\frac{\partial}{\partial x}(\phi a_x) + \frac{\partial}{\partial y}(\phi a_y) + \frac{\partial}{\partial z}(\phi a_z) = \phi\left(\frac{\partial a_x}{\partial x} + \frac{\partial a_y}{\partial y} + \frac{\partial a_z}{\partial z}\right) + \left(a_x\frac{\partial\phi}{\partial x} + a_y\frac{\partial\phi}{\partial y} + a_z\frac{\partial\phi}{\partial z}\right)$$

(2) $\operatorname{div}[ab] = -(\mathbf{a}\operatorname{curl}\mathbf{b}) + (\mathbf{b}\operatorname{curl}\mathbf{a}).$

$$\frac{\partial}{\partial x}(a_y b_z - a_z b_y) + \frac{\partial}{\partial y}(a_z b_x - a_x b_z) + \frac{\partial}{\partial z}(a_x b_y - a_y b_x)$$

$$= -a_x\left(\frac{\partial b_z}{\partial y} - \frac{\partial b_y}{\partial z}\right) - a_y\left(\frac{\partial b_x}{\partial z} - \frac{\partial b_z}{\partial x}\right) - a_z\left(\frac{\partial b_y}{\partial x} - \frac{\partial b_x}{\partial y}\right)$$

$$+ b_x\left(\frac{\partial a_z}{\partial y} - \frac{\partial a_y}{\partial z}\right) + b_y\left(\frac{\partial a_x}{\partial z} - \frac{\partial a_z}{\partial x}\right) + b_z\left(\frac{\partial a_y}{\partial x} - \frac{\partial a_x}{\partial y}\right).$$

(3) $\operatorname{curl}[ab] = \mathbf{a}\operatorname{div}\mathbf{b} - \mathbf{b}\operatorname{div}\mathbf{a} - (\mathbf{a}\nabla)\,\mathbf{b} + (\mathbf{b}\nabla)\,\mathbf{a}.$

$$\frac{\partial}{\partial y}(a_x b_y - a_y b_x) - \frac{\partial}{\partial z}(a_z b_x - a_x b_z)$$

$$= a_x\left(\frac{\partial b_x}{\partial x} + \frac{\partial b_y}{\partial y} + \frac{\partial b_z}{\partial z}\right) - b_x\left(\frac{\partial a_x}{\partial x} + \frac{\partial a_y}{\partial y} + \frac{\partial a_z}{\partial z}\right)$$

$$-\left(a_x\frac{\partial b_x}{\partial x} + a_y\frac{\partial b_x}{\partial y} + a_z\frac{\partial b_x}{\partial z}\right) + \left(b_x\frac{\partial a_x}{\partial x} + b_y\frac{\partial a_x}{\partial y} + b_z\frac{\partial a_x}{\partial z}\right).$$

(4) From (β) *if* $\mathbf{a}$ *is a constant vector*

$$\operatorname{div}[ab] = -(\mathbf{a}\operatorname{curl}\mathbf{b}).$$

If $\mathbf{a} = (1, 0, 0)$ we have the particular case (see footnote, p. 77)

$$\frac{\partial b_y}{\partial z} - \frac{\partial b_z}{\partial y} = \operatorname{div}(0,\ -b_z,\ b_y).$$

PART II

MINKOWSKI'S FOUR-DIMENSION WORLD

For the convenience of readers who desire as far as possible to avoid the following out of mathematical analysis, the proofs of statements are in the following chapters printed in smaller type. By passing over such passages the general line of argument may be followed. The sections indicated by an asterisk might also be passed over by those desirous only of following out the points peculiar to the principle of relativity. The general notions of the extended vectors as given on pp. 90–96 are however employed in Part III, and should therefore be read, at least as far as they are in the larger type.

CHAPTER VIII

MINKOWSKI'S FOUR-DIMENSION CALCULUS

1. We have now concluded the account of the manner in which the principle of relativity originated, and given some idea of the point of view which it adopts as to the nature of the measures of space and time, and of other physical magnitudes, and we may now proceed to the further development of the principle, the relations which it suggests, and the light which it throws on the significance of the earlier electromagnetic and mechanical theory. A great impetus to the work on this part of the subject was given by Minkowski. The form which is assumed by a principle which is developed to meet a special problem is generally moulded by the conditions of the problem, but it often happens that it is afterwards seen

that the form so obtained is a particular case of a more general and often more elegant principle. Minkowski, with his keen geometrical insight, was the first to notice the general mathematical significance of the Einstein kinematics.

2. The fundamental postulate of that mode of thought is the *invariance of the velocity of light*. This may be simply expressed mathematically as follows.

Let (x, y, z, t), (x', y', z', t') be any two sets of space-time coordinates consistent with this postulate.

If a point P is moving with velocity c in the first system, and $(\delta x, \delta y, \delta z)$ is its displacement in time δt, then

$$\delta x^2 + \delta y^2 + \delta z^2 - c^2 \delta t^2 = 0.$$

If in the same way $(\delta x', \delta y', \delta z')$ is the displacement of the same point as seen in the other set of coordinates in time $\delta t'$ the fundamental postulate requires that

$$\delta x'^2 + \delta y'^2 + \delta z'^2 - c^2 \delta t'^2 = 0$$

shall also be satisfied. This relation is to hold whenever the same relation holds in the first system of coordinates.

But by differentiating the functional relations connecting (x, y, z, t) with (x', y', z', t') it is seen at once that

$$\delta x'^2 + \delta y'^2 + \delta z'^2 - c^2 \delta t'^2$$

is a homogeneous quadratic function of $(\delta x, \delta y, \delta z, \delta t)$. We must therefore have

$$\delta x'^2 + \delta y'^2 + \delta z'^2 - c^2 \delta t'^2 \equiv \phi(x, y, z, t)(\delta x^2 + \delta y^2 + \delta z^2 - c^2 \delta t^2) \ldots (1),$$

where ϕ is some function of (x, y, z, t) alone—not involving the differentials—this relation being true whatever the ratios $\delta x : \delta y : \delta z : \delta t$.

If we consider a similar relation in three variables

$$\delta x'^2 + \delta y'^2 + \delta z'^2 = \phi(x, y, z)(\delta x^2 + \delta y^2 + \delta z^2),$$

and assume it satisfied for all directions of the displacement $(\delta x, \delta y, \delta z)$, this implies that the change in coordinates is what is known as a conformal transformation, that is, that the ratio of corresponding short elements of length in the neighbourhood of corresponding points is the same for all directions, so that the *shape* of any *small* element of volume is unaltered, the change in *size* being determined by the function ϕ which is the square of the linear magnification. With this in view, Minkowski introduces a new variable

$$u = ict, \text{ where } i = \sqrt{(-1)},$$

and writing the relation (1) in the form

$$\delta x'^2 + \delta y'^2 + \delta z'^2 + \delta u'^2 = \phi(x, y, z, u)(\delta x^2 + \delta y^2 + \delta z^2 + \delta u^2)$$

interprets this geometrically as indicating that the transformation is conformal in the imaginary space of four dimensions in which the coordinates are (x, y, z, u).

3. The classification of all possible transformations.

The introduction of such a mind-stuff as this into a physical theory may seem at first to make it rather elusive, but the gain in symmetry makes it possible to anticipate relations which would otherwise probably have remained undiscovered. Further it enables us to investigate all the possible types of transformation for which the fundamental postulate can be satisfied. For as in three dimensions, so in any greater number of dimensions, it has been proved that all possible conformal transformations fall into three classes*.

(i) *Transformations in which* $\phi \equiv 1$.

In three dimensions these would be the transformations corresponding to a change from any one set of axes of (x, y, z) to any other set, leaving the configuration of the system observed unaltered; or again they would express the changes in the coordinates of any point of a rigid body in the most general

* See Darboux, *Leçons sur les systèmes orthogonaux et les coordonnées curvilignes*, 2nd ed., Vol. I, § 91, p. 159. For some developments of the general transformation as an extension of the Principle of Relativity see Hassé, *Lond. Math. Soc. Proceedings*, Dec. 12, 1912.

possible displacement, the only condition satisfied being that which expresses that all lengths are unchanged.

Thus we may say that all the transformations of this group correspond to the displacements of a figure of permanent configuration. A purely *translational* displacement corresponds merely to a change of origin without rotation of the axes, and clearly is of no importance to consider here. But the most general rotational displacement depends on six parameters (e.g. the twelve coordinates of three points A, B, C connected by six relations expressing that their distances from one another and from a fixed point are unchanged). The equations of transformation are all linear, and the successive application of two transformations of this class is equivalent to a third single transformation; they constitute what is technically called a *group*.

We may note at once that the Lorentz-Einstein transformation falls into this class, for we may express it in the form

$$x' = x\cos\theta + u\sin\theta,$$

$$u' = -x\sin\theta + u\cos\theta,$$

$$y' = y, \quad z' = z,$$

where θ is the imaginary angle given by

$$\tan\theta = \frac{iv}{c}, \quad \cos\theta = \frac{1}{(1 - v^2/c^2)^{\frac{1}{2}}}, \quad \sin\theta = \frac{2v/c}{(1 - v^2/c^2)^{\frac{1}{2}}}.$$

It is clearly a generalization of the rotation of a rigid body about a fixed axis.

But there are two other classes of transformation, viz.

(ii) *Transformations in which ϕ is a constant other than unity*; these transformations are merely those of class (i) with each of the four variables multiplied by a constant. It is clear that a change in scale of both space and time will not affect the velocity of any moving point, provided the increase of unit is the same in both space and time.

(iii) The third class of transformations is a very large one, but it may be looked upon as built up from elementary transformations each of which is a *generalized inversion in a four-dimension hypersphere**.

In three dimensions the corresponding elementary transformation is an inversion in an arbitrary sphere.

Taking the centre of inversion at the origin the elementary transformation would be of the form

$$x' = \frac{k^2 x}{r^2}, \quad y' = \frac{k^2 y}{r^2}, \quad z' = \frac{k^2 z}{r^2}, \quad u' = \frac{k^2 u}{r^2},$$

where $$r^2 = x^2 + y^2 + z^2 + u^2.$$

It is easily calculated that this gives

$$\delta x'^2 + \delta y'^2 + \delta z'^2 + \delta u'^2 = \frac{k^4}{r^4}(\delta x^2 + \delta y^2 + \delta z^2 + \delta u^2).$$

It can be shewn that the invariant properties of the electromagnetic equations established for the Einstein transformation are equally valid for this transformation†.

4. Relativity and Rotation.

The question is often asked whether the new Principle of Relativity can be extended so as to indicate a possibility that a rotational motion of the whole physical universe relative to a possible frame of reference would be concealed, a possibility not admitted in Newtonian theory. In other words given one frame of reference for which the customary laws of electrodynamics are valid, is it possible to find another such that a body at rest in the first is in rotational motion in the second.

This question can now be answered decisively *in the negative.* For the general space-time transformation of the third group—the first two groups lead to nothing but uniform translational motions and changes of scale—is given by purely algebraic relations between the old and new coordinates, and no such

* The number of arbitrary parameters contained in the equations giving the simplest transformation of the group is five, four being the coordinates of the centre of inversion, and the fifth being the radius of inversion.

† *Lond. Math. Soc. Proceedings*, Feb. 11, 1909.

relation can express anything corresponding to a rotational motion of the space-frame of reference. Thus we do not appear to be brought any nearer to the removal of the old-time difficulty that the physical laws which seem best to describe the phenomena of motion postulate an absolute standard of direction though not of position, while apart from the physical phenomena there is no independent means of identifying such a direction. The point of view which has been developed above, treating the measures and standards of space and time as inseparable from the phenomena which they are used to describe, is emphasized by the persistence of the contradiction between the empirical laws and the hypothesis of an independent metaphysical space to which the conception of a fixed direction is as alien as that of an absolute velocity.

5. The Minkowski Transformations—Vectors in four-dimension space.

The discussion of Minkowski is restricted to the transformations of the first type.

If we think for a moment of the corresponding transformations in ordinary three-dimension space, which correspond to the changes in the coordinates of a point, due to a rotational displacement of the axes in any way about the origin, we remember that the equations which express the new coordinates (x', y', z') in terms of the old coordinates are of exactly the same form as those which express the components (X', Y', Z') of any *vector* quantity relative to the new axes in terms of its components relative to the old.

So in the four-dimension space of Minkowski, *any set of four quantities* (X, Y, Z, U) *in* S, *which transforms into* (X', Y', Z', U') *in* S' *in such a way that the equations connecting* (X, Y, Z, U) *with* (X', Y', Z', U') *are identical with those connecting* (x, y, z, u) *with* (x', y', z', u'), *whatever the system* S', *is said to be a 'vector quantity,'* but in order to distinguish it from another kind of quantity which has six components, it will be called here a '4-vector.'

The form of the equations of transformation is

$$x' = a_{11}x + a_{12}y + a_{13}z + a_{14}u,$$
$$y' = a_{21}x + a_{22}y + a_{23}z + a_{24}u,$$
$$z' = a_{31}x + a_{32}y + a_{33}z + a_{34}u,$$
$$u' = a_{41}x + a_{42}y + a_{43}z + a_{44}u,$$

where the coefficients a_{rs} are limited only by the condition that all distances are unaltered; that is

$$(x_2' - x_1')^2 + (y_2' - y_1')^2 + (z_2' - z_1')^2 + (u_2' - u_1')^2$$

is identically equal to

$$(x_2 - x_1)^2 + (y_2 - y_1)^2 + (z_2 - z_1)^2 + (u_2 - u_1)^2.$$

If we put $$A_{rs} = a_{1r}a_{1s} + a_{2r}a_{2s} + a_{3r}a_{3s} + a_{4r}a_{4s}$$

the necessary and sufficient conditions for this are

$$A_{rr} = 1 \quad \ldots \quad \ldots \quad \ldots \quad r = (1, 2, 3, 4),$$
$$A_{rs} = 0 \quad \ldots \quad \ldots \quad \ldots \quad r, s = (1, 2, 3, 4), \ r \neq s.$$

An important property of this type of transformation is that,

$$(x_1 y_1 z_1 u_1),\ (x_2 y_2 z_2 u_2),\ (x_3 y_3 z_3 u_3),\ (x_4 y_4 z_4 u_4)$$

being any four 4-vectors, the determinant

$$\begin{vmatrix} x_1, & y_1, & z_1, & u_1 \\ x_2, & y_2, & z_2, & u_2 \\ x_3, & y_3, & z_3, & u_3 \\ x_4, & y_4, & z_4, & u_4 \end{vmatrix}$$

is invariant.

This is easily proved algebraically. We will only point out here that it is the generalization of the known fact in three dimensions, that if a change of axes is made, no alteration is made in any volume, and therefore

$$\frac{1}{6}\begin{vmatrix} x_1, & y_1, & z_1 \\ x_2, & y_2, & z_2 \\ x_3, & y_3, & z_3 \end{vmatrix}$$

which is the volume of a tetrahedron whose angular points are

$$(x_1, y_1, z_1)\,(x_2, y_2, z_2)\,(x_3, y_3, z_3)\,(0, 0, 0)$$

is invariant.

6. The Lorentz transformation a particular case.

The equations

$$x' = \beta x + (\beta iv/c)\,u, \quad u' = \beta u - (\beta iv/c)\,x,$$
$$y' = y, \quad z' = z$$

satisfy the above requirements and these form the Lorentz transformation.

If any 4-vector has components (X, Y, Z, icT), *it is subject to the transformation*

$$X' = \beta(X - vT), \quad Y' = Y, \quad Z' = Z, \quad T' = \beta(T - vX/c^2).$$

7. The **scalar product** of two 4-vectors is defined thus.

If $\mathfrak{x} = (x_1, x_2, x_3, x_4)$ and $\mathfrak{y} = (y_1, y_2, y_3, y_4)$ the scalar product of $\mathfrak{x}$ and $\mathfrak{y}$ is

$$x_1y_1 + x_2y_2 + x_3y_3 + x_4y_4,$$

and is denoted by $(\mathfrak{x}\mathfrak{y})$.

The scalar product of two 4-vectors is invariant.

For let $\mathfrak{x}'$, $\mathfrak{y}'$ be the transformed vectors (x_1', x_2', x_3', x_4'), (y_1', y_2', y_3', y_4'). Then the product $(\mathfrak{x}'\mathfrak{y}')$ is identically equal to

$$A_{11}x_1y_1 + A_{12}(x_1y_2 + x_2y_1) + \ldots$$

and, as above, $A_{rr} = 1, \; A_{rs} = 0,$

so that

$$(\mathfrak{x}'\mathfrak{y}') \equiv x_1y_1 + x_2y_2 + x_3y_3 + x_4y_4$$
$$= (\mathfrak{x}\mathfrak{y}).$$

Conversely if any set of four quantities (X_1, X_2, X_3, X_4) *gives an invariant product when multiplied by an arbitrary 4-vector, it must constitute a 4-vector.*

For if (X_1, X_2, X_3, X_4) transforms into (X_1', X_2', X_3', X_4') and we multiply by the 4-vector $(1, 0, 0, 0)$ which transforms into $(a_{11}, a_{21}, a_{31}, a_{41})$ we have

$$X_1 = a_{11}X_1' + a_{21}X_2' + a_{31}X_3' + a_{41}X_4',$$

since the product is to be the same in both cases; and similarly, multiplying by $(0, 1, 0, 0)$ which transforms into $(a_{12}, a_{22}, a_{32}, a_{42})$,

$$X_2 = a_{12}X_1' + a_{22}X_2' + a_{32}X_3' + a_{42}X_4',$$

and so on.

Thus the given fact of the invariance of the product of (X_1, X_2, X_3, X_4) with an arbitrary 4-vector determines the coefficients in the transformation of this set of quantities.

Further since we know that any 4-vector does satisfy the condition, these coefficients will be the same as the coefficients of transformation of any 4-vector, that is (X_1, X_2, X_3, X_4) *is a 4-vector.*

8. On 6-vectors.

There is a second kind of vector with which we must deal in four-dimension space which is subject to the same type of transformation as an element of a 'two-dimensionality' or 'surface.'

If we think of the coordinates of a point on such a surface as functions of two parameters λ and μ, then in the neighbourhood of a pair of values which are not singularities we have

$$\delta x = a_1 \delta\lambda + b_1 \delta\mu,$$
$$\delta y = a_2 \delta\lambda + b_2 \delta\mu,$$
$$\delta z = a_3 \delta\lambda + b_3 \delta\mu,$$
$$\delta u = a_4 \delta\lambda + b_4 \delta\mu.$$

Taking a small closed region of values of λ and μ we obtain a small element of a 'surface.' We define six quantities $(s_1, s_2, s_3, s_4, s_5, s_6)$ by the equations

$$s_1 = \sigma \frac{\partial(y, z)}{\partial(\lambda, \mu)} = \sigma\,(a_2 b_3 - a_3 b_2),$$
$$s_2 = \sigma \frac{\partial(z, x)}{\partial(\lambda, \mu)} = \sigma\,(a_3 b_1 - a_1 b_3),$$
$$s_3 = \sigma \frac{\partial(x, y)}{\partial(\lambda, \mu)} = \sigma\,(a_1 b_2 - a_2 b_1),$$
$$s_4 = \sigma \frac{\partial(x, u)}{\partial(\lambda, \mu)} = \sigma\,(a_1 b_4 - a_4 b_1),$$
$$s_5 = \sigma \frac{\partial(y, u)}{\partial(\lambda, \mu)} = \sigma\,(a_2 b_4 - a_4 b_2),$$
$$s_6 = \sigma \frac{\partial(z, u)}{\partial(\lambda, \mu)} = \sigma\,(a_3 b_4 - a_4 b_3),$$

where if λ, μ are thought of as coordinates in a plane σ is the small area in this plane which contains all the values of λ, μ in question.

The set of six quantities $(s_1, s_2, s_3, s_4, s_5, s_6)$ will be denoted in the aggregate by $\mathfrak{S}$.

We define the *product of two such quantities* $\mathfrak{S}$, $\mathfrak{S}'$ as

$$(\mathfrak{S}\mathfrak{S}') = s_1 s_1' + s_2 s_2' + s_3 s_3' + s_4 s_4' + s_5 s_5' + s_6 s_6'.$$

By actual multiplication we find

$$(\mathfrak{S}\mathfrak{S}') = \sigma\sigma' \{\Sigma(aa') \, . \, \Sigma(bb') - \Sigma(ab') \, . \, \Sigma(a'b)\}.$$

Now the quantities

$$(a_1\delta\lambda,\ a_2\delta\lambda,\ a_3\delta\lambda,\ a_4\delta\lambda),$$
$$(b_1\delta\mu,\ b_2\delta\mu,\ b_3\delta\mu,\ b_4\delta\mu),$$

and the same quantities with dashes, constitute four 4-vectors, so that

$$\Sigma aa',\ \Sigma bb',\ \Sigma ab',\ \Sigma a'b$$

are invariants of any Minkowski transformation. Hence $(\mathfrak{S}\mathfrak{S}')$ is an invariant.

It follows as in the case of 4-vectors that all quantities of the type of $\mathfrak{S}$ undergo the same linear transformation when (x, y, z, u) are subjected to one of the Minkowski transformations.

We therefore introduce the definition—*all quantities which are subject to the same transformation as that which applies to the elements of surfaces* $\mathfrak{S}$ *will be called* 6-*vectors.*

We have the characteristic property that *the necessary and sufficient condition that a set of six quantities denoted in the aggregate by* $\mathfrak{P}$ *shall form a 6-vector is that the product of* $\mathfrak{P}$ *and an arbitrary element* $\mathfrak{S}$ *shall be an invariant.*

This is proved exactly as for 4-vectors.

9. The vector product of two 4-vectors is defined thus.

If (x_1, x_2, x_3, x_4), (y_1, y_2, y_3, y_4) are two 4-vectors $\mathfrak{x}$, $\mathfrak{y}$ then the *vector product of* $\mathfrak{x}$ *and* $\mathfrak{y}$ *is the set of six quantities*

$$(x_2y_3 - x_3y_2),\quad (x_3y_1 - x_1y_3),\quad (x_1y_2 - x_2y_1),$$
$$(x_1y_4 - x_4y_1),\quad (x_2y_4 - x_4y_2),\quad (x_3y_4 - x_4y_3),$$

and is denoted by $[\mathfrak{x}, \mathfrak{y}]$.

It is clear from the above that

$$[\mathfrak{x}, \mathfrak{y}] \textit{ is a 6-vector.}$$

N.B. The *general* 6-vector is not capable of representation by an area $\mathfrak{S}$ for the six components of $\mathfrak{S}$ are connected by the relation

$$s_1s_4 + s_2s_5 + s_3s_6 = 0,$$

whereas from the definition given of a 6-vector this is clearly not necessary.

10. The product of a 6-vector and a 4-vector.

Let $(\overline{yz}, \overline{zx}, \overline{xy}, \overline{xu}, \overline{yu}, \overline{zu})$ be any 6-vector and let

$$\overline{yz} = -\overline{zy}, \text{ etc.}$$

Then *the set of four quantities*

$$\begin{aligned} X &= \qquad\qquad\; y\,.\,\overline{xy} + z\,.\,\overline{xz} + u\,.\,\overline{xu}, \\ Y &= x\,.\,\overline{yx} \qquad\quad\;\; + z\,.\,\overline{yz} + u\,.\,yu, \\ Z &= x\,.\,\overline{zx} + y\,.\,\overline{zy} \qquad\quad\;\; + u\,.\,\overline{zu}, \\ U &= x\,.\,\overline{ux} + y\,.\,\overline{uy} + z\,.\,\overline{uz}, \end{aligned}$$

where (x, y, z, u) is any 4-vector, make up a 4-vector.

For if we form the product of (X, Y, Z, U) and any other 4-vector (x', y', z', u') we obtain

$$\begin{aligned} Xx' + Yy' + Zz' + Uu' & \\ &= \overline{yz}\,(zy' - yz') + \overline{zx}\,(xz' - zx') + \overline{xy}\,(yx' - xy') \\ &\quad + \overline{xu}\,(ux' - xu') + \overline{yu}\,(uy' - yu') + \overline{zu}\,(uz' - zu'), \end{aligned}$$

and this last expression is the product of two 6-vectors and is therefore invariant whatever the 4-vector (x', y', z', u'). Thus (X, Y, Z, U) *is a 4-vector.*

If $\mathfrak{F}$ is the 6-vector, and $\mathfrak{r} = (x, y, z, u)$, we shall denote this 4-vector by $[\mathfrak{r}\mathfrak{F}]$.

11. Reciprocal 6-vector.

With any 6-vector

$$\mathfrak{F} = (\overline{yz}, \overline{zx}, \overline{xy}, \overline{xu}, yu, \overline{zu})$$

is associated another *reciprocal* 6-*vector*, obtained by interchange of the components in pairs thus

$$\mathfrak{F}' = (\overline{xu}, \overline{yu}, \overline{zu}, \overline{yz}, \overline{zx}, \overline{xy}).$$

To shew that $\mathfrak{F}'$ is a 6-vector it is sufficient to shew that, $\mathfrak{p}$, $\mathfrak{q}$ being any 4-vectors, so that $[\mathfrak{p}\mathfrak{q}]$ is a 6-vector, $([\mathfrak{p}, \mathfrak{q}], \mathfrak{F}')$ is invariant.

Let $[\mathfrak{p}, \mathfrak{q}]'$ be the set of quantities obtained from $[\mathfrak{p}, \mathfrak{q}]$ by the same interchange as that which gives $\mathfrak{F}'$ from $\mathfrak{F}$.

Then $$([\mathfrak{p}, \mathfrak{q}], \mathfrak{F}') \equiv ([\mathfrak{p}, \mathfrak{q}]', \mathfrak{F}).$$

If therefore it can be shewn that $[\mathfrak{p}, \mathfrak{q}]'$ is a 6-vector, this product is invariant, which is what we seek to prove.

To shew that $[\mathbf{p}, \mathbf{q}]$ is a 6-vector we multiply by the 6-vector $[\mathbf{r}, \mathbf{s}]$ the product of any two other 4-vectors. Then

$$([\mathbf{pq}]' \cdot [\mathbf{rs}]) = (p_x q_u - p_u q_x)(r_y q_z - r_z q_y) + \ldots + \ldots$$
$$+ \ldots + \ldots + \ldots$$
$$= \begin{vmatrix} p_x & p_y & p_z & p_u \\ q_x & q_y & q_z & q_u \\ r_x & r_y & r_z & r_u \\ s_x & s_y & s_z & s_u \end{vmatrix}$$

and this determinant is known to be invariant (p. 91).

12. Differentiation of vectors with respect to a parameter.

If $\mathbf{r}$ is a 4-vector (x, y, z, u) which is a function of a number of parameters $(\lambda, \mu, \nu, \ldots)$, then since

$$\mathbf{r}(\lambda, \mu, \nu, \ldots) \text{ is a 4-vector}$$

and
$$\mathbf{r}(\lambda + \delta\lambda, \mu + \delta\mu, \ldots) \text{ is a 4-vector,}$$

therefore the difference $\delta\lambda \frac{\partial \mathbf{r}}{\partial \lambda}$ is a 4-vector.

Thus $\frac{\partial \mathbf{r}}{\partial \lambda}, \frac{\partial \mathbf{r}}{\partial \mu}$, etc., are all 4-vectors.

Hence *the vector product*

$$\left[\frac{\partial \mathbf{r}}{\partial \lambda}, \frac{\partial \mathbf{r}}{\partial \mu}\right],$$

that is the set of six Jacobians

$$\left(\frac{\partial (y, z)}{\partial (\lambda, \mu)}, \frac{\partial (z, x)}{\partial (\lambda, \mu)}, \frac{\partial (x, y)}{\partial (\lambda, \mu)}, \frac{\partial (x, u)}{\partial (\lambda, \mu)}, \frac{\partial (y, u)}{\partial (\lambda, \mu)}, \frac{\partial (z, u)}{\partial (\lambda, \mu)}\right),$$

is a 6-vector.

Forming the product of this 6-vector with the 4-vector $\frac{\partial \mathbf{r}}{\partial \nu}$,

$$\left[\frac{\partial \mathbf{r}}{\partial \nu}\left(\frac{\partial \mathbf{r}}{\partial \lambda}, \frac{\partial \mathbf{r}}{\partial \mu}\right)\right]$$

is a 4-vector, that is

$$\left(\frac{\partial (y, z, u)}{\partial (\lambda, \mu, \nu)}, \frac{\partial (z, u, x)}{\partial (\lambda, \mu, \nu)}, \frac{\partial (u, x, y)}{\partial (\lambda, \mu, \nu)}, \frac{\partial (x, y, z)}{\partial (\lambda, \mu, \nu)}\right)$$

constitute a 4-vector.

Lastly, multiplying this by the 4-vector $\frac{\partial \mathbf{r}}{\partial \tau}$, we have

$$\frac{\partial (x, y, z, u)}{\partial (\lambda, \mu, \nu, \tau)} \textit{ is an invariant.}$$

13. Einstein's addition equation in vectorial form.

In the four-dimensional world of Minkowski, the motion of a point in ordinary space is represented by a single curve or continuous one-dimensional aggregate of points (x, y, z, u) (Weltlinie).

If we take two points on this curve near to one another the differences of their coordinates are the components of a 4-vector; let us say

$$\delta\mathfrak{r} = (\delta x, \delta y, \delta z, \delta u).$$

Then since the absolute value of a 4-vector is invariant

$$|\delta\mathfrak{r}| = (\delta x^2 + \delta y^2 + \delta z^2 + \delta u^2)^{\frac{1}{2}} = \delta u\,(1 - w^2/c^2)^{\frac{1}{2}} = \delta u/\kappa$$

is an invariant, where w is what we call the velocity of the point in ordinary space, that is, $(\delta x^2 + \delta y^2 + \delta z^2)^{\frac{1}{2}}/\delta t$. Hence dividing the 4-vector $\delta\mathfrak{r}$ by this invariant we have

$$\kappa\left(\frac{dx}{du}, \frac{dy}{du}, \frac{dz}{du}, 1\right) \textit{ is a 4-vector},$$

or on multiplying by ic

$$\mathfrak{w} = \kappa\,(w_x, w_y, w_z, ic) \textit{ is a 4-vector}.$$

This statement gives the Einstein addition equation if we apply the particular Lorentz transformation. For, writing in full, we have for that case, as in § 6, p. 91,

$$\kappa' w_x' = \beta\left(\kappa w_x + \frac{iv}{c}\,.\,\kappa ic\right) \qquad \text{where } \beta = \left(1 - \frac{v^2}{c^2}\right)^{-\frac{1}{2}}$$

$$= \beta\kappa\,(w_x - v),$$

$$\kappa' w_y' = \kappa w_y, \quad \kappa' w_z' = \kappa w_z,$$

$$\kappa' ic = \beta\left(\kappa ic - \frac{iv}{c}\,\kappa w_x\right).$$

The last equation gives

$$\kappa' = \kappa\beta\left(1 - \frac{v w_x}{c^2}\right).$$

Substituting this in the remaining equations, we have

$$w_x' = \frac{w_x - v}{1 - \dfrac{v w_x}{c^2}},$$

$$w_y' = \frac{w_y}{\beta\left(1 - \frac{vw_x}{c^2}\right)},$$

$$w_z' = \frac{w_z}{\beta\left(1 - \frac{vw_x}{c^2}\right)}.$$

14. An important invariant.

We have seen that $\frac{\partial (x, y, z, u)}{\partial (\lambda, \mu, \nu, \tau)}$ is an invariant.

Hence *for corresponding four-dimensional regions*

$$\iiiint dx\, dy\, dz\, du = \iiiint \frac{\partial (x, y, z, u)}{\partial (\lambda, \mu, \nu, \tau)} d\lambda\, d\mu\, d\nu\, d\tau$$

is an invariant.

If the region of integration corresponds to all the points of a three-dimensional volume V in one system of coordinates, considered for a definite time T, then in any other system the region corresponds to the points of another volume, which we will call V', and corresponding elements of time are so related that $\delta t/\kappa$ is invariant.

Hence $$\iiiint dx\, dy\, dz\, dt = VT = V'T'$$

where $$\frac{T}{\kappa} = \frac{T'}{\kappa'}.$$

Thus $$\kappa' V' = \kappa V$$

or, what is the same thing, κV is invariant.

This is a generalization of Einstein's treatment of the Lorentz contraction in volume.

GENERAL THEOREMS IN FOUR-DIMENSIONAL VECTOR-ANALYSIS†

*15. The generalization of Stokes' Theorem.

The importance of the well-known theorem of Stokes‡ in

† Largely developed by Sommerfeld, *Annalen der Physik*, 32 (1910), p. 749 and 33 (1910), p. 649.

‡ That is,

$$\int \mathbf{P ds} = \int (X dx + Y dy + Z dz)$$
$$= \iint \left\{ l\left(\frac{\partial Z}{\partial y} - \frac{\partial Y}{\partial z}\right) + m\left(\frac{\partial X}{\partial z} - \frac{\partial Z}{\partial x}\right) + n\left(\frac{\partial Y}{\partial x} - \frac{\partial X}{\partial y}\right)\right\} dS$$
$$= \iint (\text{curl}\, \mathbf{P} \,.\, \mathbf{dS}).$$

mathematical physics lies in the fact that it enables us at once to see that, if $\mathbf{P} = (X, Y, Z)$, the vector 'curl,' whose components are

$$\frac{\partial Z}{\partial y} - \frac{\partial Y}{\partial z}, \quad \frac{\partial X}{\partial z} - \frac{\partial Z}{\partial x}, \quad \frac{\partial Y}{\partial x} - \frac{\partial X}{\partial y},$$

is an invariant vector; that is, if the axes be changed and the components of the vector (X, Y, Z) relative to the new axes are called (X', Y', Z'), then the quantities

$$\frac{\partial Z'}{\partial y'} - \frac{\partial Y'}{\partial z'}, \quad \frac{\partial X'}{\partial z'} - \frac{\partial Z'}{\partial x'}, \quad \frac{\partial Y'}{\partial x'} - \frac{\partial X'}{\partial y'}$$

are the components relative to the new axes of the same vector as that which is given by the former expressions relative to the old axes.

In fact the expression $\int X dx + Y dy + Z dz$ is independent of the axes used, being equal to $\int (\mathbf{P} \mathbf{ds})$. Hence $\iint (\text{curl}\, \mathbf{P} \,.\, \mathbf{dS})$ is unchanged by a change of axes whatever the area of integration, so that it follows at once that $(\text{curl}\, \mathbf{P} \,.\, \mathbf{dS})$ is an invariant whatever be the direction of the element.

We can therefore write down at once the transformation formulae for the quantity curl $\mathbf{P}$; they will be identical with the formulae for any other vector.

In the extension of this to four-dimension vectors, the invariance which follows from attaching a geometrical meaning to a vector product has to be replaced by a corresponding invariance which is demonstrated algebraically as above*.

If $\mathfrak{p} = (X, Y, Z, U)$, $\delta\mathfrak{s} = (\delta x, \delta y, \delta z, \delta u)$ *are two 4-vectors, then*

$$\int (\mathfrak{p}\mathfrak{d}\mathfrak{s}) = \int (X dx + Y dy + Z dz + U du)$$

round any closed curve is invariant.

* See pp. 92 and 93. Sommerfeld uses a quasi-geometrical language, but the appeal to pure geometry of four dimensions is in reality only disguising an algebraic argument.

This integral may be transformed into

$$\iint \left(\frac{\partial Z}{\partial y} - \frac{\partial Y}{\partial z}\right) dy\,dz + \left(\frac{\partial X}{\partial z} - \frac{\partial Z}{\partial x}\right) dz\,dx + \left(\frac{\partial Y}{\partial x} - \frac{\partial X}{\partial y}\right) dx\,dy$$
$$+ \left(\frac{\partial U}{\partial x} - \frac{\partial X}{\partial u}\right) dx\,du + \left(\frac{\partial U}{\partial y} - \frac{\partial Y}{\partial u}\right) dy\,du + \left(\frac{\partial U}{\partial z} - \frac{\partial Z}{\partial u}\right) dz\,du,$$

taken over a closed two-dimensional region of which the original curve is the boundary.

The integral is an invariant whatever be the region of integration, and in particular for any infinitesimal region round a given point whatever the direction of the element. Since therefore the elementary region of integration is itself a 6-vector, *the set of six quantities*

$$\left(\frac{\partial Z}{\partial y} - \frac{\partial Y}{\partial z}, \quad \frac{\partial X}{\partial z} - \frac{\partial Z}{\partial x}, \quad \frac{\partial Y}{\partial x} - \frac{\partial X}{\partial y},\right.$$
$$\left.\frac{\partial U}{\partial x} - \frac{\partial X}{\partial u}, \quad \frac{\partial U}{\partial y} - \frac{\partial Y}{\partial u}, \quad \frac{\partial U}{\partial z} - \frac{\partial Z}{\partial u}\right)$$

constitutes a 6-vector (p. 93). We denote it by 'curl $\mathfrak{p}$.'

*16. The generalization of Gauss' Theorem.

The last paragraph treated of an extension of Stokes' theorem. Consider now a generalization of Gauss' theorem, viz.

$$\iint X\,dy\,dz + Y\,dz\,dx + Z\,dx\,dy = \iiint \left(\frac{\partial X}{\partial x} + \frac{\partial Y}{\partial y} + \frac{\partial Z}{\partial z}\right) dx\,dy\,dz,$$

or in vector form

$$\iint (\mathbf{P}d\mathbf{S}) = \iiint \operatorname{div} \mathbf{P}\,dV.$$

We have, by exactly similar processes, the result

$$\iiint (X\,dy\,dz\,du + Y\,dz\,du\,dx + Z\,du\,dx\,dy + U\,dx\,dy\,dz)$$
$$= \iiiint \left(\frac{\partial X}{\partial x} + \frac{\partial Y}{\partial y} + \frac{\partial Z}{\partial z} + \frac{\partial U}{\partial u}\right) dx\,dy\,dz\,du.$$

The right-hand side is a scalar quantity, and $dx\,dy\,dz\,du$ (i.e. $\dfrac{\partial(x, y, z, u)}{\partial(\lambda, \mu, \nu, \tau)} d\lambda\,d\mu\,d\nu\,d\tau$) is an invariant of the transformations.

Thus taking a very small region of integration, and supposing (X, Y, Z, U) to be a 4-vector, since $(dy\,dz\,du, dz\,du\,dx, du\,dx\,dy, dx\,dy\,dz)$† is also a 4-vector, it follows that

$$\frac{\partial X}{\partial x}+\frac{\partial Y}{\partial y}+\frac{\partial Z}{\partial z}+\frac{\partial U}{\partial u}$$

is an invariant quantity. We may call it the 'divergence of the 4-vector' (X, Y, Z, U) and denote it by 'div $\mathfrak{p}$.'

*17. Conversion of a double integral into a triple integral.

Occupying an intermediate place between the theorems of Gauss and Stokes is the formula connecting a double integral with a triple integral in four dimensions.

Starting from any 6-vector $\mathfrak{P}$ and integrating over the boundary of any closed three-dimensional region we have

$$\iint(\mathfrak{P}\mathbf{d}\mathfrak{S})=\iiint q_x dy\,dz\,du+q_y dz\,du\,dx+q_z du\,dx\,dy+q_u dx\,dy\,dz,$$

where

$$q_x=\qquad\frac{\partial P_{xy}}{\partial y}+\frac{\partial P_{xz}}{\partial z}+\frac{\partial P_{xu}}{\partial u},$$

$$q_y=\frac{\partial P_{yx}}{\partial x}\qquad+\frac{\partial P_{yz}}{\partial z}+\frac{\partial P_{yu}}{\partial u},$$

$$q_z=\frac{\partial P_{zx}}{\partial x}+\frac{\partial P_{zy}}{\partial y}\qquad+\frac{\partial P_{zu}}{\partial u},$$

$$q_u=\frac{\partial P_{ux}}{\partial x}+\frac{\partial P_{uy}}{\partial y}+\frac{\partial P_{uz}}{\partial z}\qquad.$$

Let $\mathbf{d}\mathfrak{V}=(dV_x, dV_y, dV_z, dV_u)$ stand for the 4-vector $(dy\,dz\,du$, etc.). Then, since $\mathfrak{P}\mathbf{d}\mathfrak{S}$ is invariant whatever the element $\mathbf{d}\mathfrak{S}$, and since

$$\iint(\mathfrak{P}\mathbf{d}\mathfrak{S})=\iiint(q_x dV_x+q_y dV_y+q_z dV_z+q_u dV_u)$$

† That is,

$$d\lambda\,d\mu\,d\nu\left(\frac{\partial(y, z, u)}{\partial(\lambda, \mu, \nu)}, \frac{\partial(z, u, x)}{\partial(\lambda, \mu, \nu)}, \frac{\partial(u, x, y)}{\partial(\lambda, \mu, \nu)}, \frac{\partial(x, y, z)}{\partial(\lambda, \mu, \nu)}\right),$$

supposing the bounding surface to be given by $\tau=$const. See p. 96.

whatever the region of integration,

$$q_x dV_x + q_y dV_y + q_z dV_z + q_u dV_u$$

is invariant whatever the 4-vector element $\mathbf{d}\mathfrak{V}$.

Hence $\mathfrak{q} = (q_x, q_y, q_z, q_u)$ *is a 4-vector.*

*Following Minkowski it will be convenient to introduce the definition**

$$\mathfrak{q} = \text{lor } \mathfrak{P}.$$

18. A generalization of Green's Theorem.

If $\mathfrak{a}$ *is a 4-vector and* $\mathfrak{B}$ *is a 6-vector*

$$\text{div}\,[\mathfrak{a}\mathfrak{B}] = (\mathfrak{B} \text{ curl } \mathfrak{a}) + (\mathfrak{a} \text{ lor } \mathfrak{B}).$$

This is at once seen on writing out the expression on the left, namely

$$\frac{\partial}{\partial x}\{a_y B_{xy} + a_z B_{xz} + a_u B_{xu}\} + \ldots + \ldots + \ldots$$

$$= a_x \left(\frac{\partial B_{yx}}{\partial x} + \frac{\partial B_{zx}}{\partial x} + \frac{\partial B_{ux}}{\partial x}\right) + \ldots + \ldots + \ldots$$

$$+ B_{xy}\left(\frac{\partial a_y}{\partial x} - \frac{\partial a_x}{\partial y}\right) + \ldots + \ldots + \ldots + \ldots + \ldots$$

$$= (\mathfrak{a} \text{ lor } \mathfrak{B}) + (\mathfrak{B} \text{ curl } \mathfrak{a}).$$

We may say that this is a generalization of

$$\text{div}\,(\phi\mathbf{f}) = \phi \text{ div } \mathbf{f} + (\mathbf{f} \,.\, \nabla\phi),$$

where ϕ is a scalar and $\mathbf{f}$ a three-dimension vector.

In particular if $\mathfrak{B} = \text{curl } \mathfrak{b}$, we have from this equation

$$\text{div}\,[\mathfrak{a} \text{ curl } \mathfrak{b}] - \mathfrak{a} \text{ lor curl } \mathfrak{b} = \text{curl } \mathfrak{a} \text{ curl } \mathfrak{b}$$

$$= \text{div}\,[\mathfrak{b} \text{ curl } \mathfrak{a}] - \mathfrak{b} \text{ lor curl } \mathfrak{a}.$$

This is analogous to the equation

$$\text{div}\,(\phi \,.\, \nabla\psi) - \phi \text{ div } (\nabla\psi) = \nabla\phi\nabla\psi$$

$$= \text{div}\,(\psi \,.\, \nabla\phi) - \psi \text{ div } (\nabla\phi).$$

* Suggested by the name of Lorentz.

CHAPTER IX

THE FIELD EQUATIONS OF THE ELECTRON THEORY IN MINKOWSKI'S FORM

1. We may now apply the above analysis to the equations of electromagnetic theory.

If we take the fundamental equations of Lorentz,

$$-\frac{1}{c}\frac{\partial \mathbf{e}}{\partial t} + \text{curl } \mathbf{h} = \frac{\rho \mathbf{u}}{c},$$

$$\text{div } \mathbf{e} = \rho,$$

$$\frac{1}{c}\frac{\partial \mathbf{h}}{\partial t} + \text{curl } \mathbf{e} = 0,$$

$$\text{div } \mathbf{h} = 0,$$

and write

$$\mathfrak{s} = (s_x, s_y, s_z, s_u) = \rho \left\{ \frac{u_x}{c}, \frac{u_y}{c}, \frac{u_z}{c}, i \right\}$$

and

$$F_{yz} = h_x, \quad F_{zx} = h_y, \quad F_{xy} = h_z,$$

$$F_{xu} = -ie_x, \quad F_{yu} = -ie_y, \quad F_{zu} = -ie_z,$$

calling the aggregate of these last six quantities $\mathfrak{F}$ the first pair of the above equations becomes

$$\frac{\partial F_{xy}}{\partial y} + \frac{\partial F_{xz}}{\partial z} + \frac{\partial F_{xu}}{\partial u} = s_x,$$

$$\frac{\partial F_{yx}}{\partial x} + \frac{\partial F_{yz}}{\partial z} + \frac{\partial F_{yu}}{\partial u} = s_y,$$

$$\frac{\partial F_{zx}}{\partial x} + \frac{\partial F_{zy}}{\partial y} + \frac{\partial F_{zu}}{\partial u} = s_z,$$

$$\frac{\partial F_{ux}}{\partial x} + \frac{\partial F_{uy}}{\partial y} + \frac{\partial F_{uz}}{\partial z} = s_u.$$

Now we have seen in the preceding chapter that if $\mathfrak{F}$ is a 6-vector the left-hand members of these equations constitute a 4-vector which has been denoted by 'lor $\mathfrak{F}$'.

Hence if we suppose that $\mathfrak{s}$ is a 4-vector the above equations represent *the equality of two 4-vectors, a relation which is invariant for any of the transformations.*

But we have seen that if

$$\mathfrak{F} = (F_{yz},\ F_{zx},\ F_{xy},\ F_{xu},\ F_{yu},\ F_{zu})$$

is a 6-vector, so also is

$$\mathfrak{F}' = (F_{xu},\ F_{yu},\ F_{zu},\ F_{yz},\ F_{zx},\ F_{xy}).$$

Hence interchanging $-i\mathbf{e}$ and $\mathbf{h}$ in $\mathfrak{F}$ we see that

$$\mathfrak{F}' = (-i\mathbf{e},\ \mathbf{h})$$

is a 6-vector.

Hence lor $\mathfrak{F}'$ is a 4-vector.

But the second pair of equations gives for the particular set of coordinates used

$$\text{lor}\ \mathfrak{F}' = 0,$$

and this expresses another invariant relation.

Thus by the use of Minkowski's notation the invariant form of the equations is expressed in the statement that *the field equations are vectorial relations, and therefore independent of the particular set of coordinates used,* and the conditions for the invariance are

(i) $\mathfrak{F} = (\mathbf{h},\ -i\mathbf{e})$ *is a 6-vector,*

(ii) $\mathfrak{s} = \rho\,(u_x,\ u_y,\ u_z,\ ic)/c$ *is a 4-vector.*

2. The total charge in an element of volume is invariant.

This follows at once from (ii).

For $\mathfrak{s} = \rho\left(\frac{u_x}{c},\ \frac{u_y}{c},\ \frac{u_z}{c},\ i\right)$ is a 4-vector.

The square of a 4-vector is invariant.

But $$\mathfrak{s}^2 = \rho^2 \left(\frac{u_x^2 + u_y^2 + u_z^2}{c^2} - 1 \right) = - \rho^2/\kappa^2.$$

Hence ρ/κ is invariant.

In fact the 4-vector $\mathfrak{s}$ and the velocity 4-vector $\kappa(u_x, u_y, u_z, ic)$ are two 4-vectors in the same direction, their invariant ratio being $\rho/\kappa c$.

But if δV is an element of volume moving with velocity (u_x, u_y, u_z) we have seen above that

$$\kappa \delta V \text{ is invariant.}$$

Multiplying these two quantities we have

$$\rho \delta V \text{ is invariant.}$$

3. Conservation of electric charge.

We have identically

$$\operatorname{div} (\operatorname{lor} \mathfrak{F}) = 0$$

hence $$\operatorname{div} \mathfrak{s} = 0.$$

This is the equation of continuity of electricity, expressing the fact that the charge moves but does not change in amount.

4. Other invariants in the most general transformation.

Since the product of two 6-vectors is an invariant, we have, in the case of the free aether,

$$(\mathbf{h}, -i\mathbf{e})^2 = \mathbf{h}^2 - \mathbf{e}^2$$

is an invariant; that is *the difference of the squares of the intensities of electric and magnetic energies is an invariant.*

Remembering too that $\dfrac{\partial (x, y, z, u)}{\partial (\lambda, \mu, \nu, \tau)}$ is invariant, $(\lambda, \mu, \nu, \tau)$ being any four independent parameters, we have also

$$\iiiint (\mathbf{h}^2 - \mathbf{e}^2)\, dx dy dz dt$$

is an invariant, for corresponding regions of integration.

This invariance associates itself at once with the derivation of the field equations of the free aether as those which are necessary to give this integral a stationary value. A knowledge of this fact would have enabled us to anticipate the invariant form of those equations. This will be further dealt with below (pp. 111–2).

Further since $\mathfrak{F}' = (-i\mathbf{e}, \mathbf{h})$ is also a 6-vector the product

$$(\mathfrak{F}\mathfrak{F}') = (\mathbf{h}, -i\mathbf{e})(-i\mathbf{e}, \mathbf{h}) = -2i(\mathbf{eh})$$

is an invariant scalar quantity.

Thus we have the invariant property that '*if the electric and magnetic intensities are perpendicular in any one system of reference, they are perpendicular in any other.*'

*5. The electrodynamic potential.

The equation $$\text{lor}\,\mathfrak{F}' = 0$$
is satisfied identically if we put

$$\mathfrak{F} = \text{curl}\,\mathfrak{a},$$

where $\mathfrak{a}$ is any 4-vector function.

We shall call this 4-vector the '*electrodynamic potential.*'

Substituting in the equation

$$\text{lor}\,\mathfrak{F} = \mathfrak{s}$$

we obtain $$\text{lor curl}\,\mathfrak{a} = \mathfrak{s}.$$

The first component of this equation expanded is

$$\frac{\partial}{\partial y}\left(\frac{\partial a_y}{\partial x} - \frac{\partial a_x}{\partial y}\right) + \frac{\partial}{\partial z}\left(\frac{\partial a_z}{\partial x} - \frac{\partial a_x}{\partial z}\right) + \frac{\partial}{\partial u}\left(\frac{\partial a_u}{\partial x} - \frac{\partial a_x}{\partial u}\right) = s_x,$$

that is $$\frac{\partial}{\partial x}(\text{div}\,\mathfrak{a}) - \Box\, a_x = s_x,$$

where $\Box\,\phi$ stands for

$$\frac{\partial^2\phi}{\partial x^2} + \frac{\partial^2\phi}{\partial y^2} + \frac{\partial^2\phi}{\partial z^2} + \frac{\partial^2\phi}{\partial u^2}.$$

We may now, if we choose, subject '$\mathfrak{a}$' to the restricting condition

$$\text{div}\,\mathfrak{a} = 0,$$

for as in three dimensions we can form a vector field having any given circulation and divergence, and the restriction is not inconsistent with the equation satisfied by $\mathfrak{a}$†. The 4-vector $\mathfrak{a}$ then satisfies the equation

$$\Box \mathfrak{a} = -\mathfrak{s},$$

which is an exact generalization of the equation satisfied by the electrostatic potential ϕ, namely $\nabla^2\phi = -\rho$.

In ordinary notation we have

$$h_x = F_{yz} = \frac{\partial a_z}{\partial y} - \frac{\partial a_y}{\partial z}, \quad h_y = \frac{\partial a_x}{\partial z} - \frac{\partial a_z}{\partial x}, \quad h_z = \frac{\partial a_y}{\partial x} - \frac{\partial a_x}{\partial y},$$

and $$-ie_x = F_{xu} = \frac{\partial a_u}{\partial x} - \frac{\partial a_x}{\partial u},$$

or $$e_x = i\frac{\partial a_u}{\partial x} - \frac{1}{c}\frac{\partial a_x}{\partial t}, \quad e_y = i\frac{\partial a_u}{\partial y} - \frac{1}{c}\frac{\partial a_y}{\partial t}, \quad e_z = i\frac{\partial a_u}{\partial z} - \frac{1}{c}\frac{\partial a_z}{\partial t}.$$

Here clearly (a_x, a_y, a_z) is Maxwell's 'vector-potential' (F, G, H), while $a_u = i\Phi$ where Φ is what is known as the 'scalar electromagnetic potential.' (See Maxwell, *Electricity and Magnetism*, § 598, 3rd ed.)

*6. Retarded potentials.

We have seen above that the electric and magnetic intensities $\mathbf{e}$, $\mathbf{h}$ of the fundamental electron theory can be expressed in terms of a 4-*vector potential* by means of the equation

$$\mathfrak{F} = (\mathbf{h}, -i\mathbf{e}) = \text{curl}\, \mathfrak{a},$$

where $\mathfrak{a}$ satisfies the equation

$$\Box \mathfrak{a} = \mathfrak{s}$$

and $$\mathfrak{s} = \rho\left(\frac{\mathbf{u}}{c}, i\right).$$

† In fact we have, since

$$\text{div}\, \mathfrak{s} = 0,$$

$$\Box\,(\text{div}\, \mathfrak{a}) - \text{div}\,(\Box\, \mathfrak{a}) = 0,$$

which is an identity.

Now if the field is due to the motion of point electrons, the potential satisfies the equation

$$\Box \mathfrak{a} = 0,$$

except at the isolated points occupied by the charges at any instant.

The following values of the ordinary vector and scalar potentials have been given by Liénard and Wiechert* for the field due to the motion of a single charge e in any manner:

$$\phi = \frac{e}{4\pi \left[r\left(1 - \frac{u_r}{c}\right)\right]}, \qquad \mathbf{a} = \frac{e}{4\pi c}\left[\frac{\mathbf{u}}{r\left(1 - \frac{u_r}{c}\right)}\right],$$

where r is the distance from the point (x, y, z) at time t to the point occupied by the electron at time $\left(t - \frac{r}{c}\right)$ and the values of $\mathbf{u}$ and u_r are to be taken for the time $\left(t - \frac{r}{c}\right)$ also.

Using the Minkowski notation we have

$$\mathfrak{a} = (\mathbf{a}, i\phi) = \frac{e}{4\pi c}\left[\frac{\mathbf{u}, ic}{r\left(1 - \frac{u_r}{c}\right)}\right].$$

If we call (x_1, y_1, z_1) the coordinates of the electron at time t_1 the condition

$$t_1 = t - \frac{r}{c}$$

gives $\qquad (x - x_1)^2 + (y - y_1)^2 + (z - z_1)^2 - c^2 (t - t_1)^2 = 0;$

and subject to this condition

$$\mathfrak{a} = \frac{e}{4\pi} \cdot \frac{(\mathbf{u}/c, i)}{c(t - t_1) - \{(x - x_1) v_x + (y - y_1) v_y + (z - z_1) v_z\}/c}$$

$$= -\frac{e}{4\pi} \frac{\mathfrak{u}}{(\mathfrak{r}\mathfrak{u})},$$

where $\mathfrak{u}$ is the 4-vector

$$\kappa\left(\frac{\mathbf{u}}{c}, i\right)_{t_1}$$

and $\mathfrak{r}$ is the 4-vector

$$\{(x - x_1,\ y - y_1,\ z - z_1,\ ic(t - t_1)\}.$$

* A. Liénard, *L'Éclairage électrique* 16 (1898), pp. 5, 53, 106; Wiechert, *Arch. Néerl.* (2), 5 (1900), p. 549.

The fact that $\mathfrak{a}$ is a 4-vector, and hence the relativity of the potentials of Liénard and Wiechert, is now clearly shewn by the fact that the factor $\mathfrak{u}$ in the numerator is a 4-vector while the factor $(\mathfrak{r}\mathfrak{u})$ in the denominator is an invariant.

In fact the above expression is the simplest 4-vector expression that can be written down subject to the condition that, for the particular case of the electrostatic field due to the charge e permanently at rest, it shall reduce to the form given by

$$\mathbf{a} = 0, \quad \phi = \frac{e}{4\pi r},$$

namely

$$\mathfrak{a} = \frac{e}{4\pi}\left(0, 0, 0, \frac{i}{r}\right)\dagger;$$

and it is only necessary to shew that this 4-vector satisfies the equation

$$\Box\,\mathfrak{a} = 0$$

at all points and instants except those for which $\mathfrak{r}$ vanishes, in order to verify that $\mathfrak{a}$ is actually the potential of the field.

*7. It may be worth while to give here a simple proof of this fact‡.

We start from the simplest solution of the equation

$$\Box\,\phi = 0,$$

namely

$$\phi = \frac{1}{\mathfrak{r}^2} = \{(x - x_1)^2 + (y - y_1)^2 + (z - z_1)^2 - c^2(t - t_1)^2\}^{-1},$$

which is an immediate extension of the ordinary solution $\phi = \frac{1}{r}$ of the equation $\nabla^2\phi = 0$.

Now in the case of an electron moving along a curve let its coordinates (x_1, y_1, z_1, t_1) be expressed as analytic functions of a single parameter σ the *real* values of which determine the sequence of positions taken up at various times.

† Cf. below the work founded on that of Poincaré for modifying the law of gravitation to conform to the Principle of Relativity.

‡ The proof here given is somewhat similar to that given by Herglotz, *Gött. Nach.* 1904, p. 549. For a very full account see Abraham, *Th. der Elek.* II. 2nd ed. (1908), pp. 35–58.

Then on substitution we have

$$\phi = f(x, y, z, t, \sigma),$$

and this expression satisfies

$$\Box \phi = 0$$

identically, and will therefore satisfy it equally if we allow σ to take an imaginary value, though it would not then correspond to an actual point (x_1, y_1, z_1, t_1).

If σ_1 is the real value of σ corresponding to a particular point (x_1, y_1, z_1, t_1) which is such that

$$(x - x_1)^2 + (y - y_1)^2 + (z - z_1)^2 = c^2 (t - t_1)^2,$$

consider $\Phi = \int \phi d\sigma$ taken round a finite closed circuit of complex values of σ enclosing σ_1. Then this quantity also satisfies $\Box \phi = 0$.

Let ϕ be expanded as a function of $(\sigma - \sigma_1)$ in the neighbourhood of $\sigma = \sigma_1$, keeping (x, y, z, t) fixed. We have

$$\begin{aligned} \phi &= \frac{1}{(\mathfrak{r} - \mathfrak{r}_1 + \delta \mathfrak{r}_1)^2} \\ &= \frac{1}{(\mathfrak{r} - \mathfrak{r}_1)^2 + 2(\mathfrak{r} - \mathfrak{r}_1, \delta \mathfrak{r}_1)} \\ &= \frac{1}{2(\mathfrak{r} - \mathfrak{r}_1, \delta \mathfrak{r}_1)} \\ &= \frac{1}{2\left(\mathfrak{r} - \mathfrak{r}_1, \dfrac{d\mathfrak{r}_1}{dt_1}\right)(\sigma - \sigma_1)\dfrac{dt_1}{d\sigma_1}} \end{aligned}$$

to the first order.

Thus ϕ becomes infinite to the first order at $\sigma = \sigma_1$ with residue

$$\left(2\{(x - x_1) u_x + (y - y_1) u_y + (z - z_1) u_z - c^2 (t - t_1)\} \frac{dt_1}{d\sigma_1}\right)^{-1}.$$

Hence

$$\begin{aligned} \Phi &= \int \phi d\sigma \\ &= \pi i \frac{d\sigma_1}{dt_1} \Big/ \{(x - x_1) u_x + (y - y_1) u_y + (z - z_1) u_z - c^2 (t - t_1)\}, \end{aligned}$$

and this is a solution of $\quad \Box \Phi = 0,$

whatever be the real parameter σ_1 used to determine the position and instant (x_1, y_1, z_1, t_1). If in turn we put $\sigma_1 = x_1, y_1, z_1$ and t_1 we get

$$\Phi = \pi i \frac{(u_x, u_y, u_z, i)}{(x-x_1)u_x + (y-y_1)u_y + (z-z_1)u_z - c^2(t-t_1)}.$$

Thus it is verified that the expression

$$-\frac{e}{4\pi}\frac{\mathfrak{u}}{(\mathfrak{r}\mathfrak{u})}$$

for $\mathfrak{a}$ is a solution of $\qquad \Box\,\mathfrak{a} = 0.$

*8. Relation of the fundamental equations to the principle of least action.

The analogy just suggested between the electrostatic equations in three dimensions, and the electrodynamic in four, may be carried further.

It is well known that for a given distribution of charge of density ρ, the vector field which makes

$$\iiint \tfrac{1}{2}\mathbf{E}^2\,dx\,dy\,dz$$

stationary, subject to

$$\operatorname{div}\mathbf{E} = \rho,$$

is that in which $\mathbf{E}$ is the gradient of a potential.

Consider now the variation of the invariant integral

$$A = \iiiint \tfrac{1}{2}\mathfrak{F}^2\,dx\,dy\,dz\,du,$$

subject to $\qquad \operatorname{lor}\mathfrak{F} = \mathfrak{s},$

where $\mathfrak{F}$ is a 6-vector, and $\mathfrak{s}$ is a 4-vector given at all points of the four-dimensional region, $\mathfrak{s}$ being kept constant in the variation.

Then $\qquad \delta A = \iiiint \mathfrak{F}\,\delta\mathfrak{F}\,dx\,dy\,dz\,du,$

and the variation is subject to

$$\operatorname{lor}\mathfrak{F} = \mathfrak{s},$$

$$\operatorname{lor}(\mathfrak{F} + \delta\mathfrak{F}) = \mathfrak{s},$$

since $\qquad \delta\mathfrak{s} = 0.$

Hence $$\text{lor}\,\delta\mathfrak{F} = 0,$$

and therefore the 6-vector $\delta\mathfrak{F}'$ reciprocal to $\delta\mathfrak{F}$ can be obtained from a vector potential

$$\delta\mathfrak{F}' = \text{curl}\,\delta\mathfrak{a}.$$

But
$$\begin{aligned}\mathfrak{F}\,\delta\mathfrak{F} &\equiv \mathfrak{F}'\,\delta\mathfrak{F}'\\ &= \mathfrak{F}'\,\text{curl}\,\delta\mathfrak{a}\\ &= \text{div}\,[\delta\mathfrak{a}\,\mathfrak{F}'] - (\delta\mathfrak{a}\,.\,\text{lor}\,\mathfrak{F}')\end{aligned}$$

by the generalization of Green's Theorem, p. 102.

Hence $$\delta A = -\iiiint \delta\mathfrak{a}\,\text{lor}\,\mathfrak{F}'\,dx\,dy\,dz\,du + I,$$

where I is a triple integral over the outer boundary of the four-dimensional region of integration.

Assuming that I vanishes when the region is the whole of space and time, we shall obtain a stationary value for A for an arbitrary variation in the vector potential $\mathfrak{a}$ if, and only if,

$$\text{lor}\,\mathfrak{F}' = 0.$$

Hence it follows*, interpreting the results in the ordinary notation, that *the condition that the 'action'*

$$\iiiint \tfrac{1}{2}\,(\mathbf{h}^2 - \mathbf{e}^2)\,dx\,dy\,dz\,dt$$

shall be stationary, subject to the equations

$$\frac{1}{c}\left(\frac{\partial \mathbf{e}}{\partial t} + \rho\mathbf{u}\right) = \text{curl}\,h,$$

$$\rho = \text{div}\,e,$$

for the most general variation of e, h subject to ρ *and u being kept constant, is that*

$$-\frac{1}{c}\frac{\partial \mathbf{h}}{\partial t} = \text{curl}\,\mathbf{e},$$

and $$0 = \text{div}\,\mathbf{h}.$$

The further question of the variation of A when ρ and $\mathbf{u}$ are varied will be considered later (pp. 158–60).

* Cf. Larmor, *Aether and Matter*, Chap. VI.

CHAPTER X

MINKOWSKI'S ELECTRODYNAMICS OF MOVING BODIES

[For a summary of the main conclusions of this chapter, see pp. 133–4.]

1. Returning to the vexed question of the electrodynamic equations of moving bodies, we have now a powerful means of attack from the point of view of the 'principle of relativity.' Leaving aside all direct experimental investigation as to what is the proper form of the equations, and also all constitutive theories of matter such as that of Lorentz, let us examine what forms of the equations are consistent with the general hypothesis of the complete relativity of all phenomena. In Minkowski's terminology this hypothesis is that *all relations shall be vectorial relations in the four-dimensional space.* The equality of two 4-vectors or 6-vectors is a relation which is independent of the choice of coordinates.

We have seen that Lorentz' fundamental equations have been put into the vectorial form

$$\text{lor}\ \mathfrak{f} = \mathfrak{s},$$

$$\text{lor}\ \mathfrak{f}' = 0,$$

where $$\mathfrak{f} = (\mathbf{h}, -i\mathbf{e}), \quad \mathfrak{f}' = (-i\mathbf{e}, \mathbf{h}),$$

and $$\mathfrak{s} = \rho\left(\frac{\mathbf{u}}{c}, i\right),$$

the necessary and sufficient conditions for the invariance of the equations being that $\mathfrak{f}$, $\mathfrak{f}'$ shall be 6-vectors and $\mathfrak{s}$ a 4-vector.

Now in all theories of the phenomena in moving bodies the equations proposed have the form

$$-\frac{1}{c}\frac{\partial \mathbf{D}}{\partial t} + \operatorname{curl} \mathbf{H} = \mathbf{S},$$

$$\operatorname{div} \mathbf{D} = \rho,$$

$$\frac{1}{c}\frac{\partial \mathbf{B}}{\partial t} + \operatorname{curl} \mathbf{E} = 0,$$

$$\operatorname{div} \mathbf{B} = 0.$$

This is the form as used by Minkowski. To obtain that of Lorentz we substitute for $\mathbf{H}$, $\mathbf{H} + [\mathbf{uP}]/c$, where $\mathbf{P} = \mathbf{D} - \mathbf{E}$; and for $\mathbf{S}$ put $(\mathbf{j} + \rho\mathbf{u})/c$, where $\mathbf{j}$ is called the *conduction* current. To obtain that of Hertz we substitute for $\mathbf{E}$, $\mathbf{E} - [\mathbf{uB}]/c$, for $\mathbf{H}$, $\mathbf{H} + [\mathbf{uD}]/c$, and for $\mathbf{S}$, $(\mathbf{j} + \rho\mathbf{u})/c$.

These equations may be written

$$\operatorname{lor} \mathfrak{F} = \mathfrak{s},$$

$$\operatorname{lor} \mathfrak{G}_1 = 0,$$

where

$$\mathfrak{F} = (\mathbf{H}, -i\mathbf{D}),$$

$$\mathfrak{G}_1 = (\mathbf{E}, i\mathbf{B}),$$

and

$$\mathfrak{s} = (\mathbf{S}, \rho i).$$

These equations will be invariant provided $\mathfrak{F}$ and $\mathfrak{G}_1$ are 6-vectors, and $\mathfrak{s}$ is a 4-vector.

The principle of relativity requires then that

$$(\mathbf{H}, -i\mathbf{D}), \quad (\mathbf{E}, i\mathbf{B}),$$

and the associated quantities

$$\mathfrak{F}_1 = (\mathbf{D}, i\mathbf{H}), \quad \mathfrak{G} = (\mathbf{B}, -i\mathbf{E}),$$

obtained by interchange as in § 11, p. 95, *and afterwards multiplying by i, shall be* 6-*vectors.*

2. These conditions may be compared with those obtained earlier as a consequence of the Lorentz constitutive theory of matter (p. 82). For the particular Einstein transformation they become

$$D_x = d_x, \quad D_y = \beta\left(d_y - \frac{v}{c}h_z\right), \quad D_z = \beta\left(d_z + \frac{v}{c}h_y\right) \quad \text{............}(\delta),$$

$$H_x=h_x,\quad H_y=\beta\left(h_y+\frac{v}{c}d_z\right),\quad H_z=\beta\left(h_z-\frac{v}{c}d_y\right)\quad\ldots\ldots\ldots\ldots(\eta),$$

$$E_x=e_x,\quad E_y=\beta\left(e_y-\frac{v}{c}b_z\right),\quad E_z=\beta\left(e_z+\frac{v}{c}b_y\right)\quad\ldots\ldots\ldots\ldots(\epsilon),$$

$$B_x=b_x,\quad B_y=\beta\left(b_y+\frac{v}{c}e_z\right),\quad B_z=\beta\left(b_z-\frac{v}{c}e_y\right)\quad\ldots\ldots\ldots\ldots(\beta).$$

If we take **h**, **d**, **e**, **b** to refer to a point at which the velocity of the matter is momentarily zero, the velocity at the point is $-v$ in the system in which the corresponding quantities are **H**, **D**, **E**, **B**.

If, in order to change to Lorentz' notation, we put $\mathbf{H}+[\mathbf{uP}]/c$, that is, $(H_x, H_y+vP_z/c, H_z-vP_y/c)$ for **H**, we have

$$H_x=h_x,\quad H_y+\frac{v}{c}P_z=\beta\left(h_y+\frac{v}{c}d_z\right),\quad H_z-\frac{v}{c}P_y=\beta\left(h_z-\frac{v}{c}d_y\right)\quad\ldots(\eta_1)$$

in place of (η).

Subtracting (η_1) from (β) and putting $\mathbf{M}=\mathbf{B}-\mathbf{H}$, $\mathbf{m}=\mathbf{b}-\mathbf{h}$,

$$M_x=m_x,\quad M_y-\frac{v}{c}P_z=\beta\left(m_y-\frac{v}{c}p_z\right),\quad M_z+\frac{v}{c}P_y=\beta\left(m_z+\frac{v}{c}p_y\right)$$
$$\ldots\ldots\ldots(\mu_1),$$

and subtracting (ϵ) from (δ)

$$P_x=p_x,\quad P_y=\beta\left(p_y+\frac{v}{c}m_z\right),\quad P_z=\beta\left(p_z-\frac{v}{c}m_y\right)\quad\ldots\ldots(\varpi),$$

whence $$M_y=\beta\left(1-\frac{v^2}{c^2}\right)m_y=m_y/\beta,\quad M_z=\beta\left(1-\frac{v^2}{c^2}\right)m_z=m_z/\beta\ldots(\mu_2),$$

which are exactly the equations obtained above (p. 82) by the kinematical method.

3. Analysis of the stream 4-vector.

If we consider the components of the 4-vector '$\mathfrak{S}$' in two systems of coordinates in one of which the point of the material medium in question is momentarily at rest, and in the other of which it is moving with velocity v parallel to the axis of x, and if we apply the Lorentz transformation for 4-vectors (the same transformation as that for x, y, z, t) we have

$$(S_x)_0=\beta\,(S_x-v\rho/c),$$

$$(S_y)_0 = S_y,$$
$$(S_z)_0 = S_z,$$
$$\rho_0 = \beta(\rho - vS_x/c),$$

where the suffix $_0$ applies to the system in which the point is at rest.

We may interpret these equations as follows. The quantity $c(S_x)_0$, being the total current at a stationary point, will be called the *conduction current* for the body at rest. If we call $cS - \rho v$ the conduction current (j) for the moving medium, the following equations hold for the transformation of the conduction current and density

$$j_x = (j_x)_0/\beta, \quad j_y = (j_y)_0, \quad j_z = (j_z)_0;$$
$$\rho = \beta\left\{\rho_0 + \frac{v}{c^2}(j_x)_0\right\},$$

which are the same equations as those obtained directly from the Einstein formulae save that the sign of v is different throughout.

More generally we may put these results as follows. Let $\mathfrak{u}$ stand for the velocity 4-vector

$$\kappa\left(\frac{u_x}{c}, \frac{u_y}{c}, \frac{u_z}{c}, i\right),$$

which is such that $\mathfrak{u}^2 = -1$. Suppose $\mathfrak{s}$ the total stream 4-vector to be resolved into two parts one parallel to $\mathfrak{u}$ and the other orthogonal to it*; that is

let $$\mathfrak{s} = \lambda\mathfrak{u} + \mathfrak{j},$$

where $(\mathfrak{j}\mathfrak{u}) = 0$, and λ is an invariant.

Then $$(\mathfrak{s}\mathfrak{u}) = \lambda\mathfrak{u}^2 + (\mathfrak{j}\mathfrak{u})$$
$$= -\lambda$$
$$= (\mathfrak{s}_0\mathfrak{u}_0).$$

* Two 4-vectors **a**, **b** are said to be orthogonal if their scalar product (**ab**) is zero.

For a body at rest the velocity 4-vector $\mathfrak{u}$ becomes $(0, 0, 0, i)$, so that, $\mathfrak{j}$ being orthogonal to it, we have

$$(\mathfrak{j}_u)_0 = 0.$$

Thus the resolution of $\mathfrak{s}$ becomes in this case

$$\mathfrak{s}_0 = (\mathfrak{j}_x, \mathfrak{j}_y, \mathfrak{j}_z, \rho_0 i).$$

Hence

$$(\mathfrak{s}_0 \mathfrak{u}_0) = -\rho_0 = -\lambda.$$

Thus $\mathfrak{s}$ is resolved into a convection current

$$\rho \mathfrak{u} = \kappa \rho_0 \left(\frac{u_x}{c}, \frac{u_y}{c}, \frac{u_z}{c}, i \right),$$

and a conduction current

$$\mathfrak{j} = \mathfrak{s} + \mathfrak{u} (\mathfrak{s} \mathfrak{u}).$$

4. The Constitutive Equations of a Material Medium*.

The electromagnetic phenomena in a material medium, besides being subject to the general equations of the field which are the same in form for all media, depend also upon the characteristic properties of the particular medium. These properties are represented by certain equations connecting the various quantities $\mathbf{E}$, $\mathbf{D}$, $\mathbf{H}$, $\mathbf{B}$, ρ, $\mathbf{j}$, the simplest examples being the equations

$$\mathbf{D} = \epsilon \mathbf{E}, \quad \mathbf{B} = \mu \mathbf{H}, \quad \mathbf{j} = \sigma \mathbf{E} \ldots\ldots (1), (2), (3),$$

which are characteristic of the simplest class of isotropic bodies at rest.

Now these 'constitutive' equations, as they are called, depend on exactly those factors in the constitution of the medium which are unaccounted for in the fundamental electron theory. They

* Mr Hassé suggests to me that the treatment given here of the constitutive equations appropriate to media in motion is of wider scope than that of Minkowski. It has been constantly stated that the Principle of Relativity has no bearing on the phenomena in bodies which are in non-uniform motion, and many writers read Minkowski's work in such a way as to limit the validity of the results to the case of uniform motion. It is hoped that the treatment given above may make it clear that while the Principle of Relativity cannot say what ARE the actual equations for bodies with variable motion, it does act as a criterion of what those equations *may be*, and gives a very powerful means of devising such equations to be tested by experiment.

are therefore empirical and approximate, and various forms are proposed to explain such properties of material media as dispersion, the rotation of the plane of polarization of light, and various other electro- and magneto-optical phenomena.

Now the principle of relativity in its generality, as we have seen, has a bearing on those factors in the constitution of matter which are not included in the fundamental differential equations. The experiments of Rayleigh and Brace, for example, suggest that the scope of the principle must extend to the whole machinery of optical transmission in transparent media. That of Trouton and Rankine involves a consideration of the conductivity of metals to electric currents.

The application of the principle of relativity to the constitutive equations involves then that *these relations shall be restricted to have an invariant form when the electromagnetic magnitudes are transformed according to the various equations that have been found necessary for maintaining the invariant form of the field equations**.

5. The three simple equations (1), (2), (3) quoted above are vector equations, and independent of the set of three-dimensional axes employed, as they should be since they express physical characteristics of the medium, if we adhere to the Newtonian relativity.

The principle of relativity requires that *any equations adopted shall be vector equations in the four-dimensional space of Minkowski; this is a sufficient criterion for the maintenance of the principle.*

This criterion cannot be expected to determine uniquely what the equations must be; it only limits our choice in adopting an empirical equation.

The materials at our disposal in making the choice are the various generalized vectors that have been developed above. It will be convenient to recall them.

* See also for the general criterion of relativity, Chapter XII, pp. 161–2.

6-vectors:

$\mathfrak{F} = (\mathbf{H}, -\iota\mathbf{D})$ and the associated 6-vector $\mathfrak{F}_1 = (\mathbf{D}, \iota\mathbf{H})$,
$\mathfrak{G} = (\mathbf{B}, -\iota\mathbf{E})$ and the associated 6-vector $\mathfrak{G}_1 = (\mathbf{E}, \iota\mathbf{B})$.

4-vectors:

$$\mathfrak{u} = \kappa\left(\frac{u_x}{c}, \frac{u_y}{c}, \frac{u_z}{c}, i\right),$$

$$\mathfrak{e} = [\mathfrak{u}\mathfrak{F}\ ] = \kappa\,\{\mathbf{E} + [\mathbf{uB}]/c,\ i\,(\mathbf{E\,u})/c\},$$
$$\mathfrak{b} = [\mathfrak{u}\mathfrak{F}_1] = \kappa\,\{\mathbf{B} - [\mathbf{uE}]/c,\ i\,(\mathbf{B\,u})/c\},$$
$$\mathfrak{d} = [\mathfrak{u}\mathfrak{G}\ ] = \kappa\,\{\mathbf{D} + [\mathbf{uH}]/c,\ i\,(\mathbf{Du})/c\},$$
$$\mathfrak{h} = [\mathfrak{u}\mathfrak{G}_1] = \kappa\,\{\mathbf{H} - [\mathbf{uD}]/c,\ i\,(\mathbf{Hu})/c\}.$$

It is now easy to write down relations connecting these quantities, which will for the case of bodies at rest lead to such three-dimensional equations as (1), (2), (3), p. 117.

Thus, for example, if we put

$$\mathfrak{d} = \epsilon\mathfrak{e} \quad \text{.........................(1)},$$

which is equivalent to

$$\mathbf{D} + [\mathbf{uH}]/c = \epsilon\,\{\mathbf{E} + [\mathbf{uB}]/c\} \quad \text{...........(i)},$$

together with the equation which follows from this

$$(\mathbf{Du}) = \epsilon\,(\mathbf{Eu}),$$

we have an equation of invariant form reducing to

$$\mathbf{D} = \epsilon\,\mathbf{E}$$

when $\mathbf{u} = 0$.

Similarly the equation

$$\mathfrak{b} = \mu\mathfrak{h} \quad \text{.........................(2)}$$

is equivalent to $\mathbf{B} - [\mathbf{uE}]/c = \mu\,\{\mathbf{H} - [\mathbf{uD}]/c\}$ ....(ii),

with the resulting equation

$$(\mathbf{Bu}) = \mu\,(\mathbf{Hu}),$$

and is an invariant equation reducing to

$$\mathbf{B} = \mu\mathbf{H}$$

when $\mathbf{u} = 0$.

Equations (i) and (ii) are therefore constitutive equations which are consistent with complete relativity, the velocity $\mathbf{u}$ of the moving matter not being restricted in any way to be constant, but only to be subject to the Einstein velocity transformation. It may be repeated however that these equations are *suggested*, not demonstrated, by the above argument.

In the same way if we desire an equation suitable to represent the facts of absorption and dispersion in a moving body we might take an equation of the form

$$a\frac{\partial^2 \mathbf{P}}{\partial t^2}+b\frac{\partial \mathbf{P}}{\partial t}+c\mathbf{P}=\mathbf{E} \quad \text{..................(4)}$$

as characteristic of a medium at rest*, and seek to obtain an equation between 4-vectors which reduces to this in the case of rest. Writing

$$\mathfrak{p}=\mathfrak{d}-\mathfrak{e}$$

we have a 4-vector which reduces for the case of rest to

$$(\mathfrak{p})_0=(P_x, P_y, P_z, 0).$$

Further, as we saw above†, if the 4-vector $\mathfrak{p}$ varies, then $\dot{\mathfrak{p}}=\kappa\frac{\partial \mathfrak{p}}{\partial t}$ is another 4-vector, and again $\ddot{\mathfrak{p}}=\kappa\frac{\partial}{\partial t}\left(\kappa\frac{\partial \mathfrak{p}}{\partial t}\right)$ is another. These results are true whether the velocity is assumed to be a constant or not.

Then since the sum of a number of 4-vectors is a 4-vector, the equation

$$a\ddot{\mathfrak{p}}+b\dot{\mathfrak{p}}+c\mathfrak{p}=\mathfrak{e} \quad \text{....................(4)}$$

would be an equation of invariant form, and so would satisfy the principle of relativity, while reducing for the case of a medium at permanent rest to the equation

$$a\frac{\partial^2 \mathbf{P}}{\partial t^2}+b\frac{\partial \mathbf{P}}{\partial t}+c\mathbf{P}=\mathbf{E}$$

as desired.

But we are not able to postulate the equation (4) except in the case of uniform velocity, because by definition the 4-vectors $\mathfrak{e}$, $\mathfrak{p}$ are subject to $(\mathfrak{e}\mathfrak{u})=0$, $(\mathfrak{p}\mathfrak{u})=0$ and this is not necessarily

* Cf. Lorentz, *Theory of Electrons*, pp. 139–40.

† p. 97.

true of $\dot{\mathfrak{p}}$ and $\ddot{\mathfrak{p}}$ unless $\mathfrak{u}$ is constant. The equations (**1**) and (**2**) however are not subject to any restriction of this sort.

6. The conduction equation.

The generalization of the equation

$$\mathbf{j} = \sigma \mathbf{E} \quad \text{.........................(3)}$$

expressing the property of a conducting medium requires a little further consideration. We are not able to write

$$\mathfrak{s} = \sigma \mathfrak{e},$$

since $(\mathfrak{e}\mathfrak{u}) = 0$ and this would involve the restriction $(\mathfrak{s}\mathfrak{u}) = 0$.

But the analysis of the total stream $\mathfrak{s}$ into a *conduction* and *convection* current (p. 117), of which the latter is orthogonal to $\mathfrak{u}$, allows us to take as a possible equation

$$\mathfrak{j} = \sigma \mathfrak{e} \quad \text{.........................(}\mathbf{3}\text{)}.$$

According to the results of p. 117 this is equivalent to

$$\rho \mathbf{u}/c + \mathbf{j} = \sigma \kappa \{\mathbf{E} + [\mathbf{uB}]/c\} + \rho_0 \kappa \mathbf{u}/c$$

and
$$\rho i = \sigma i \kappa (\mathbf{uE})/c + \rho_0 \kappa i,$$

whence
$$\mathbf{j} = \beta \sigma \{\mathbf{E} + [\mathbf{uB}]/c - \mathbf{u}(\mathbf{Eu})/c^2\} \text{...........(iii)}.$$

For $\mathbf{u} = (v, 0, 0)$, this gives $j_x = \sigma E_x/\beta$, $j_y = \sigma\beta\{E_y - vB_z/c\}$, $j_z = \sigma\beta\{E_z + vB_y/c\}$.

Thus the conductivity of the moving medium, defined as the ratio of the component current in any direction to the component of the force $(\mathbf{E} + [\mathbf{uB}]/c)$ in that direction, is *not the same for all directions*, but for the direction of motion is less than the conductivity when the medium is at rest in the ratio $1:(1 - v^2/c^2)^{\frac{1}{2}}$ and for directions at right angles is increased in the same ratio.

As in the case of the other constitutive equations we could in a similar manner devise generalizations of any given relation between $\mathbf{j}$ and $\mathbf{E}$ to fit with the principle of relativity.

7. The Invariants in the Equations for Material Media.

From the 6-vectors

$$\mathfrak{F} = (\mathbf{H}, -\iota\mathbf{D}),$$
$$\mathfrak{F}_1 = (\mathbf{D}, \iota\mathbf{H}),$$
$$\mathfrak{G} = (\mathbf{B}, -\iota\mathbf{E}),$$
$$\mathfrak{G}_1 = (\mathbf{E}, \iota\mathbf{B}),$$

we can form the invariant products

$$\mathfrak{F}^2 = \mathbf{H}^2 - \mathbf{D}^2,$$
$$(\mathfrak{F}\mathfrak{F}_1) = 2\mathbf{DH},$$
$$\mathfrak{G}^2 = \mathbf{B}^2 - \mathbf{E}^2,$$
$$(\mathfrak{G}\mathfrak{G}_1) = 2\mathbf{EB},$$
$$\mathfrak{F}\mathfrak{G} = \mathfrak{F}_1\mathfrak{G}_1 = \mathbf{BH} - \mathbf{ED},$$
$$\mathfrak{F}\mathfrak{G}_1 = \mathfrak{F}_1\mathfrak{G} = \mathbf{BD} + \mathbf{EH}.$$

The total charge in a given element of volume is *not* invariant, as in the case of the electron theory equations, except when there is no conduction current. This is not a contradiction of the assumption of the conservation of change, but has been shewn in Part I* to follow directly from that assumption combined with the Einstein kinematics.

8. Faraday's Law of Induction for moving circuits.

It is well known that the equations

$$-\frac{1}{c}\frac{\partial \mathbf{B}}{\partial t} = \operatorname{curl} \mathbf{E},$$
$$\operatorname{div} \mathbf{B} = 0,$$

lead to the equation

$$\frac{1}{c}\frac{\mathbf{D}}{\partial t}\iint \mathbf{B}d\mathbf{S} = \int (\mathbf{E} + [\mathbf{uB}]/c, \, d\mathbf{s}),$$

where the symbol $\frac{\mathbf{D}}{\partial t}$ is used to indicate that the closed area of

* See p. 83.

integration is bounded by a moving curve, the velocity of the element ds being u, and that the operations of differentiation and integration are not interchangeable.

We have seen (p. 114) how to bring the fundamental equations into line with the principle of relativity by writing them in the invariant form

$$\text{lor}\,\mathfrak{G}_1 = 0.$$

It follows that *the law of induction will maintain its form when* **E** *and* **B** *are transformed in the proper manner*, that is in the manner implied by saying that $\mathfrak{G}_1 = (\mathbf{E}, i\mathbf{B})$ is a 6-vector.

It is probably worth while here to give a deduction of the law of induction from the fundamental equations in the form used by Minkowski.

The equation $\text{lor}\,\mathfrak{G}_1 = 0$

is equivalent to the equation

$$\iint \mathfrak{G}_1 \delta \mathfrak{S} = 0$$

taken over the boundary of any three-dimensional region in the space (x, y, z, ict).

Written in full this equation is

$$0 = \iint E_x dx\,dt + E_y dy\,dt + E_z dz\,dt - (B_x dy\,dz + B_y dz\,dx + B_z dx\,dy)/c.$$

Now let the closed two-dimensional region of integration be defined as follows: take an unclosed surface S in ordinary (x, y, z) space at a fixed time t, its boundary being called s. Suppose the points of the surface A to move, the velocity of a typical point being represented by u, and let A' be the displaced position of the surface at time $t + \delta t$.

The points of s in moving to their new position on A' trace out a two-dimension locus B (in x, y, z, t) which together with A and A' forms a closed region.

If now we divide the whole integral into parts taken over A, A' and B respectively we can proceed thus. On A and A', t does not vary. Thus for these two surfaces we obtain the integral

$$\delta t \frac{d}{dt} \iint (B_x dy\,dz + B_y dz\,dx + B_z dx\,dy)/c,$$

the differentiation referring to the moving surface.

On B the coordinates of any point are functions of two parameters, viz. a (the arc of s) and t. Expressing the various terms of the integral in terms of these parameters, we have

$$\iint E_x dx\,dt = \iint E_x \frac{\partial(x, t)}{\partial(a, t)} da\,dt$$
$$= \iint E_x \frac{\partial(x_0 + u_x t, t)}{\partial(a, t)} da\,dt$$
$$= \iint E_x \frac{\partial x_0}{\partial a} da\,dt,$$

neglecting integrals such as

$$\iint E_x \frac{\partial u_x}{\partial a} t\,da\,dt,$$

which will be of the order δt^2.

Thus to this order

$$\iint_B dx\,dt = \delta t \int_s E_x dx_0$$

and again

$$\iint_B B_x dy\,dz = \iint B_x \frac{\partial(y, z)}{\partial(a, t)} da\,dt$$
$$= \iint B_x \frac{\partial(y_0 + u_y t,\ z_0 + u_z t)}{\partial(a, t)} da\,dt$$
$$= \iint B_x \begin{vmatrix} \dfrac{\partial y_0}{\partial a}, & \dfrac{\partial z_0}{\partial a} \\ u_y, & u_z \end{vmatrix} da\,dt,$$

neglecting as before terms of order δt^2.

Thus
$$\iint B_x dy\,dz = \delta t \int_s B_x (u_z dy_0 - u_y dz_0).$$

Thus taking all terms together the original integral is equal to

$$\delta t \left[\int_a dx_0 \{E_x + (B_z u_y - B_y u_z)/c\} + dy_0 \{\quad\} + dz_0 \{\quad\} \right.$$
$$\left. + \frac{1}{c}\frac{d}{dt} \iint_A (B_x dy\,dz + B_y dz\,dx + B_z dx\,dy). \right.$$

Thus we are led to the equation

$$\int_s (\mathbf{E} + [\mathbf{uB}]/c,\ ds) + \frac{1}{c}\frac{d}{dt}\int_A \mathbf{B}d\mathbf{S} = 0.$$

In the same way that the law of induction for moving circuits is obtained, we may from the equations

$$\frac{1}{c}\left(\frac{\partial \mathbf{D}}{\partial t} + \mathbf{S}\right) = \text{curl}\,\mathbf{H},$$
$$\text{div}\,\mathbf{D} = \rho,$$

obtain

$$\frac{D}{\partial t}\iint (\mathbf{D}\,.\,d\mathbf{S}) = \int (\mathbf{H} - [\mathbf{uD}]/c,\, d\mathbf{s}) - \iint (\mathbf{S} - \rho\mathbf{u})\, d\mathbf{S}$$

$$= \int \mathbf{H}' d\mathbf{s} - \iint \mathbf{j} d\mathbf{S},$$

which is the generalization to a moving circuit of the ordinary circuital equation for stationary circuits.

9. Boundary conditions at the surface of a moving medium.

These follow from the fundamental equations of the field which are assumed to hold everywhere, the discontinuity at the boundary surface of two material media being indicated by an abrupt change in the form of the constitutive equations. The case where there is on the boundary surface a surface layer of charge or current must always be considered as a limiting case of a thin finite volume distribution of charge or current, but in practice it is an important one, so that the equations will be obtained for the general case in which such a surface layer is present.

(1) The two equations

$$\operatorname{div} \mathbf{B} = 0,$$
$$\operatorname{div} \mathbf{D} = \rho,$$

lead in the ordinary way to the accepted equations

$$\mathbf{B}_{n_1} + \mathbf{B}_{n_2} = 0,$$
$$\mathbf{D}_{n_1} + \mathbf{D}_{n_2} = \sigma,$$

where the suffixes n_1, n_2 indicate the components along the normal away from the dividing surface in each case, and σ is the surface density of charge on that surface.

(2) Apply now Faraday's law of induction to a small area A which is bounded in the usual way by two elements of adjacent normals PQ, RS to the surface joined by tangent lines QR, SP, this area being supposed to move with the surface.

Let $PQ = RS = \delta n$, $QR = SP = \delta s$; let E_{s_1}' denote the component of E' along QR, and E_{s_2}' the component along SP, E_{n_1}' along PQ and E_{n_2}' along RS. Then going round the circuit in the sense $PQRS$ we have

$$\int \mathbf{E}'d\mathbf{s} = (E_{s_1}' + E_{s_2}')\,\delta s + (E_{n_1}' + E_{n_2}')\,\delta n.$$

Let B_n be the component of $\mathbf{B}$ perpendicular to the area A; this is continuous, and will be assumed to have a finite differential coefficient. It will also be assumed that $\mathbf{u}$ the velocity of either medium is continuous in the neighbourhood of the boundary. With these assumptions we may write

$$\frac{d}{dt}\iint \mathbf{B}d\mathbf{S} = bA = b\,\delta s\,\delta n,$$

where b is a *finite* quantity.

Dividing by δs we have

$$E_{s_1}' + E_{s_2}' = -\frac{1}{c}\,b\delta n - (E_{n_1}' + E_{n_2}')\,\frac{\delta n}{\delta s}.$$

Pass now to the limiting case by making δn and δs both tend to zero in an arbitrarily small ratio, and we obtain

$$E_{s_1}' + E_{s_2}' = 0.$$

Since E_{s_1}' and E_{s_2}' are in opposite directions this means that the *tangential component of* $\mathbf{E}'$, *that is, of* $\mathbf{E} + [\mathbf{uB}]/c$, *is continuous.*

If we take the corresponding circuital equation for H' and proceed in the same way, we obtain

$$\int \mathbf{H}'d\mathbf{a} = (H_{s_1}' + H_{s_2}')\,\delta s + (H_{n_1}' + H_{n_2}')\,\delta n,$$

$$\frac{D}{\partial t}\iint \mathbf{D}d\mathbf{A} = d\,\delta s\,\delta n,$$

$$\int \mathbf{j}d\mathbf{A} = k,$$

where d is finite and k is the total flow *by conduction* through the circuit.

Thus dividing by δs and proceeding to the limit as before

$$H_{s_1}' + H_{s_2}' = \underset{\delta s \to 0}{\mathrm{Lt}} \frac{k}{\delta s}.$$

Now if the current density j is everywhere finite, $\frac{k}{\delta s \delta n}$ is finite, and therefore

$$\underset{\delta s \to 0}{\mathrm{Lt}} \frac{k}{\delta n} = 0.$$

But if there is a current sheet in the dividing surface, $\mathrm{Lt}\,(k/\delta s)$ may be finite, and this limit is the current per unit length across δs. Calling this κ we have

$$H_{s_1}' + H_{s_2}' = \kappa,$$

remembering that the sense of κ is such that a right-handed rotation about it gives the direction of s_1 on one side of the surface and s_2 on the other.

If $\kappa = 0$ then H_{s_1}', *that is, the tangential component of* $\mathbf{H} - [\mathbf{uD}]/c$ *is continuous; or, in the notation of Lorentz,* $\mathbf{H} + [\mathbf{uP}]/c - [\mathbf{uD}]/c$, *that is,* $\mathbf{H} - [\mathbf{uE}]/c$, *is continuous as to its tangential component.*

10. Comparison of the Minkowski constitutive equations with other theories and with experiment.

The attempts that have been made in the past to formulate constitutive equations for *moving* media consistent with the results of experiment do not extend beyond a generalization of the simplest cases of the equations for bodies at rest, viz.

$$\mathbf{D} = \epsilon\mathbf{E}, \quad \mathbf{B} = \mu\mathbf{H}, \quad \mathbf{j} = \sigma\mathbf{E}.$$

The forms which are suggested by the application of the principle of relativity to replace these have been obtained above (pp. 119–21). They are, in Minkowski's notation, if u is the velocity of the medium,

$$\mathbf{D} + [\mathbf{uH}]/c = \epsilon\,\{\mathbf{E} + [\mathbf{uB}]/c\} \quad \text{......................(i)},$$

$$\mathbf{B} - [\mathbf{uE}]/c = \mu\,\{\mathbf{H} - [\mathbf{uD}]/c\} \quad \text{....................(ii)},$$

$$\mathbf{j} = \beta\sigma\,\{\mathbf{E} + [\mathbf{uB}]/c - \mathbf{u}\,(\mathbf{Eu})/c^2\} \quad \text{...(iii)}.$$

If we neglect $(u/c)^2$ we obtain from (iii) the equation

$$\mathbf{j} = \sigma(\mathbf{E} + [\mathbf{uB}]/c),$$

which is the equation that has been adopted by Lorentz.

The only experiment that bears on this constitutive equation is that of Trouton and Rankine which, as has been described, resulted decisively in a null effect to the order of $(u/c)^2$ and therefore, in supporting the general hypothesis of relativity to this order, confirms the equation as the appropriate modification of the empirical equation of Lorentz.

To compare equations (i) and (ii) with those proposed by Lorentz we must make the change to his notation by substituting $\mathbf{H} + [\mathbf{u}, \mathbf{D} - \mathbf{E}]/c$ for $\mathbf{H}$.

Doing this we have

$$\text{(i}\,a) \quad \mathbf{D} + [\mathbf{uH}]/c + [\mathbf{u}\,[\mathbf{u}, (\mathbf{D} - \mathbf{E})/c]]/c = \epsilon(\mathbf{E} + [\mathbf{uB}]/c),$$

$$\text{(ii}\,a) \quad \mathbf{B} - [\mathbf{uE}]/c = \mu(\mathbf{H} - [\mathbf{uE}]/c).$$

The first of these equations becomes on neglecting $(u/c)^2$

$$\text{(i}\,b) \quad \mathbf{D} = \epsilon(\mathbf{E} + [\mathbf{uB}]/c) - [\mathbf{uH}]/c,$$

while the second is exactly

$$\text{(ii}\,b) \quad \mathbf{B} = \mu(\mathbf{H} - [\mathbf{uE}]/c) + [\mathbf{uE}]/c.$$

Writing $\mathbf{M} = \mathbf{B} - \mathbf{H}$ and $\mathbf{P} = \mathbf{D} - \mathbf{E}$ as usual, substituting for $\mathbf{B}$ from (ii b) in (i b), and again neglecting $(u/c)^2$, we get

$$\text{(i}\,c) \quad \mathbf{P} = (\epsilon - 1)\,\mathbf{E} + (\epsilon\mu - 1)\,[\mathbf{uB}]/c,$$

$$\text{(ii}\,c) \quad \mathbf{M} = (\mu - 1)(\mathbf{H} - [\mathbf{uE}]/c).$$

For the particular case of a medium in which $\mu = 1$ we have

$$\text{(i}\,d) \quad \mathbf{P} = (\epsilon - 1)(\mathbf{E} + [\mathbf{uB}]/c),$$

an equation which has been verified by Wilson experimentally and adopted provisionally by Lorentz.

It has so far been impossible to verify the term involving $\mathbf{u}$ in (ii c) owing to the difficulty of obtaining a non-conducting medium in which μ is appreciably different from unity, so that the term $[\mathbf{uE}]/c$ is illusory owing to the vanishing of the electric force. We see therefore that to the order to which experiment has anything to say as to the correctness of Lorentz' theory as represented by the fundamental field equations, together with the empirical constitutive equations, it is entirely in accord with the principle of relativity.

11. Hertz' Theory.

In the same way by the substitution for $\mathbf{E}$ of $\mathbf{E} - [\mathbf{uB}]/c$, for $\mathbf{H}$ of $\mathbf{H} + [\mathbf{uD}]/c$ and for $\mathbf{S}$ of $(\mathbf{j} + \rho\mathbf{u})/c$ (see p. 114) we may obtain constitutive equations which render the field equations of Hertz consistent with the principle of relativity*.

If this is done the equations (i), (ii), (iii) become, to the first order in (u/c),

(i *e*) $$\mathbf{D} + [\mathbf{uH}]/c = \epsilon\mathbf{E},$$

(ii *e*) $$\mathbf{B} - [\mathbf{uE}]/c = \mu\mathbf{H},$$

(iii *e*) $$\mathbf{j} = \sigma\mathbf{E}.$$

It is the difference between these equations and the equations (i), (ii), (iii), which Hertz assumed to be true whether the body was moving or at rest, that must be tested by the results of experiment before we can say definitely whether it is possible or not to reconcile the field equations of Hertz with the Principle of Relativity. It is clear that in comparing with experiment the interpretation of the meaning of the vectors $\mathbf{E}$, $\mathbf{D}$, $\mathbf{B}$, $\mathbf{H}$ in the various theories must not be left ambiguous. It is by means of the boundary conditions obtained above that we are enabled to pass from the interior of the moving medium, where the meanings of the symbols are difficult of definition, to the free space through which it moves, in which the symbols must in every case have their ordinary significance.

* It is often taken for granted that the field equations of Hertz are by themselves inconsistent with the principle of relativity. It should be emphasized that this is not so, except in so far as they are united with a special group of constitutive relations connecting them.

12. Wilson's experiment*.

The object of this experiment was to examine the effect on a dielectric body of motion through a magnetic field and especially to test the discrepancy between the equations (i *b*) and (i *e*). It was found that when this dielectric moved at right angles to the direction of a magnetic field of intensity **H** with velocity **u** there was an apparent electric polarization in the direction at right angles to **u** and **H**, and approximately proportional to the product; that is apparently an equation of the form

$$\mathbf{D} = \epsilon\mathbf{E} + k[\mathbf{uH}]$$

was satisfied, reducing to the ordinary equation $\mathbf{D} = \epsilon\mathbf{E}$ when $\mathbf{u} = 0$, k being a constant to be determined.

Let the dielectric body be supposed to be a plate with its faces parallel to the plane of **u** and **H**, lying between two metal plates parallel to the same plane, and connected to the terminals

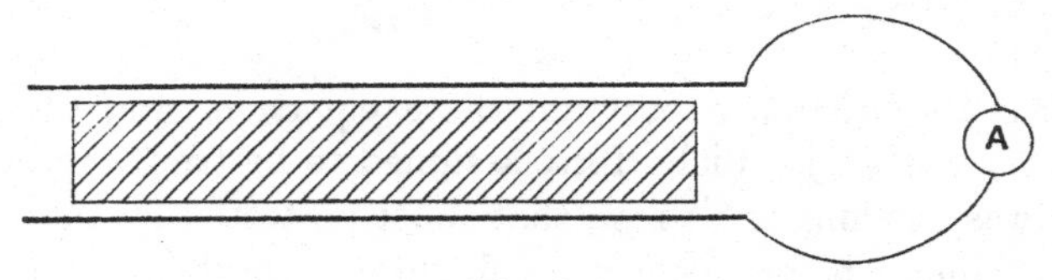

Fig. 6.

of an electrometer A. If ϕ is the difference of potential, and **E** the intensity between the plates, we have, neglecting the distance between the dielectric and the metal plates,

$$\phi = Ed,$$

where d is the thickness; and again, if σ is the surface density on the metal plate,

$$\mathbf{D} = \sigma,$$

and

$$\sigma = C\phi,$$

where C is a constant depending on the area of the metal plates and the capacity of the electrometer.

Thus we have $$\phi\left(C - \frac{\epsilon}{d}\right) = k[\mathbf{uH}].$$

* H. A. Wilson, *Phil. Trans.* (A), 204 (1904), p. 121.

The values of C, ϵ, d, $\mathbf{u}$ and $\mathbf{H}$ being known, the measurement of ϕ affords a means of determining k.

This represents, simplified from experimental details, the method of Wilson's experiment; the conclusion arrived at was that

$$k = \epsilon - 1.$$

In discussing Wilson's experiment from the point of view of Minkowski's equations, it is to be remembered that, since the field is stationary, $\frac{\partial \mathbf{B}}{\partial t} = 0$, so that curl $\mathbf{E} = 0$ and $\mathbf{E}$ is therefore the gradient of a potential, whether inside or outside the moving matter, this potential being continuous at the surface of the medium.

To use Hertz' field equations we have seen that we must put $\mathbf{E}' = \mathbf{E} - [\mathbf{uB}]/c$ for $\mathbf{E}$, and the difference of potential between any two points will then be $\int(\mathbf{E}'/c, \mathbf{ds})$. If this is done and the equation (i e) is adopted as the constitutive equation, we can write, remembering that we are assuming $\mu = 1$,

$$\mathbf{D} = \epsilon\mathbf{E}' + (\epsilon - 1)[\mathbf{uH}]/c,$$

where $\int \mathbf{E}'\mathbf{ds} = \phi$, and there is no essential difference between the calculation and that following Lorentz, and the agreement with Wilson's experiment would exist. In fact the whole difference between this modification of Hertz' equations and the equations of Lorentz would be one of notation, and definition of what is meant by the 'electric force' within a moving body. The vector $\mathbf{E} + [\mathbf{uB}]/c$ of Lorentz must be identical with the vector $\mathbf{E}$ of Hertz.

But if Hertz' original equation (i) is adhered to we have, while maintaining the equation $\int \mathbf{E}'\mathbf{ds} = \phi$,

$$\mathbf{D} = \epsilon\mathbf{E}' + \epsilon[\mathbf{uH}]/c,$$

so that any effect dependent on the last term would be greater than that predicted by the equation (i e) and confirmed by Wilson, in the ratio $\epsilon : \epsilon - 1$.

Thus Hertz' original theory, including the equation $\mathbf{D} = \epsilon\mathbf{E}$, is inconsistent with the result of this experiment.

13. Eichenwald's experiment*.

The object of this experiment was the investigation of the magnetic field produced by a polarized dielectric in motion. A circular disc of the dielectric substance, coated with a metallic sheet on each face so as to form a condenser, was rotated rapidly about an axis perpendicular to it, the two faces being kept at different electric potentials so as to produce an electric field between them. It was then found that a magnetic field was produced in the neighbourhood of the condenser which, as measured by a magnetometer, was proportional to the angular velocity and to the difference of potential between the plates, but which was independent of the nature of the dielectric.

This result is generally interpreted thus. The density of the charge at any point on the metal plates of the condenser is for a given difference of potential proportional to the specific inductive capacity ϵ of the dielectric, and therefore produces at a given point if in motion with velocity v a magnetic force proportional to ϵVv. But owing to the polarization of the dielectric there is an 'apparent' density at the contiguous point of the surface of the dielectric proportional to $-(\epsilon - 1)V$, giving rise to a magnetic force proportional to $-(\epsilon - 1)Vv$ to be added to the above force ϵVv, so that the total force is proportional to Vv and independent of ϵ.

It may be useful however to look at the result in its bearing on the principle of relativity. The rotary motion in the experiment is the only convenient means of communicating a sufficiently large measurable velocity to the dielectric. The result is usually taken as being true for any state of motion. Here we shall only consider the case of uniform rectilinear motion as a whole. This being so we may choose a system of reference in which the condenser is at rest. Then at any particular point outside the condenser, the distance between the plates of the condenser being small compared with their diameters, the electrostatic field is clearly proportional to the difference of potential 'V' between the plates, and does not depend on the nature of the dielectric. This follows from the

* A. Eichenwald, *Ann. d. Phys.* 11 (1904), p. 421.

fact that the plates of the condenser form almost a closed surface over which the potential is known, and in terms of which the exterior field may be determined without reference to what is between the plates. The magnetic field is zero in this frame of reference since the system is at rest. Let the electric field be (E_x, E_y, E_z). Then changing to a system of reference in which the system has a velocity $(-v, 0, 0)$, we find from equations (α), p. 33, that the magnetic field is

$$\mathbf{H}' = \beta\left(0, \frac{v}{c}E_z, -\frac{v}{c}E_y\right)$$
$$= \beta\,[\mathbf{vE}]/c,$$

and therefore, since $\mathbf{E}$ is proportional to V and is independent of ϵ, $\mathbf{H}'$ is to the first order in v/c proportional to Vv and also independent of ϵ, in accordance with the experimental result.

Summary of the Conclusions of this Chapter

Constitutive Equations.

The sets of field equations for moving material bodies as differently proposed by Lorentz and Hertz both fall under the type

$$-\frac{1}{c}\frac{\partial \mathbf{D}}{\partial t} + \operatorname{curl}\mathbf{H} = \mathbf{S}$$
$$\operatorname{div}\mathbf{D} = \rho,$$
$$\frac{1}{c}\frac{\partial \mathbf{B}}{\partial t} + \operatorname{curl}\mathbf{E} = 0$$
$$\operatorname{div}\mathbf{B} = 0,$$

the essential difference being one of notation rather than form. In order to make a *complete* scheme of equations these 'field equations' have to be supplemented by 'constitutive equations' characteristic of the matter in question. With the notation used above, that of Minkowski, a set of equations consonant with the principle of relativity and with the requirements of existing experimental knowledge is

$$\mathbf{D} + [\mathbf{uH}]/c = \epsilon\,\{\mathbf{E} + [\mathbf{uB}]/c\} \quad \text{(i)},$$
$$\mathbf{B} - [\mathbf{uE}]/c = \mu\,\{\mathbf{H} - [\mathbf{uD}]/c\} \quad \text{(ii)},$$
$$\mathbf{S} + \beta^2\mathbf{u}\,\{(\mathbf{Su}) - \rho c^2\}/c^2 = \sigma\beta\,\{\mathbf{E} + [\mathbf{uB}]/c\} \quad \text{(iii)}.$$

Translated into the notation of Lorentz these constitutive equations become to the first order in u/c

$$\mathbf{D} = \epsilon\,(\mathbf{E} + [\mathbf{uB}]/c) - [\mathbf{uH}]/c \;\ldots\ldots\ldots\ldots \text{(i } b\text{)},$$

$$\mathbf{B} = \mu\,(\mathbf{H} - [\mathbf{uE}]/c) + [\mathbf{uE}]/c \ldots\ldots\ldots\ldots \text{(ii } b\text{)},$$

$$\mathbf{j} = \sigma\,(\mathbf{E} + [\mathbf{uB}]/c) \;\ldots\ldots\ldots\ldots\ldots\ldots\ldots \text{(iii } b\text{)},$$

and in the notation of Hertz, they become

$$\mathbf{D} = \epsilon\mathbf{E} - [\mathbf{uH}]/c \;\ldots\ldots\ldots\ldots\ldots\ldots \text{(i } e\text{)},$$

$$\mathbf{B} = \mu\mathbf{H} + [\mathbf{uE}]/c \;\ldots\ldots\ldots\ldots\ldots\ldots \text{(ii } e\text{)},$$

$$\mathbf{j} = \sigma\mathbf{E} \;\ldots\ldots\ldots\ldots\ldots\ldots\ldots\ldots\ldots\ldots \text{(iii } e\text{)}.$$

The constitutive equations adopted by Lorentz, viz.

$$\mathbf{D} = \epsilon\mathbf{E} + (\epsilon - 1)\,[\mathbf{uH}]/c,$$

$$\mathbf{B} = \mu\mathbf{H},$$

$$\mathbf{j} = \sigma\,(\mathbf{E} + [\mathbf{uB}]/c),$$

agree with (i b, ii b, iii b) to the order contemplated in experiments hitherto performed, μ being unity in the experiment of Wilson, but the equations adopted by Hertz, viz.

$$\mathbf{D} = \epsilon\mathbf{E},$$

$$\mathbf{B} = \mu\mathbf{H},$$

$$\mathbf{j} = \sigma\mathbf{E},$$

differ from (i e, ii e, iii e) by a term of the first order in the first of the three equations and are not reconcilable with the experiments of Wilson and Eichenwald.

Boundary Conditions.

In *Minkowski's* notation, the conditions satisfied at a surface of discontinuity in the material media are that the following should be continuous:

the tangential components of $\mathbf{E} + [\mathbf{uB}]/c$ and $\mathbf{H} - [\mathbf{uD}]/c$
and the normal components of $\mathbf{B}$ and $\mathbf{D}$.

In *Lorentz'* notation the corresponding conditions are the continuity of

the tangential components of $\mathbf{E} + [\mathbf{uB}]/c$ and $\mathbf{H} - [\mathbf{uE}]/c$
and of the normal components of $\mathbf{B}$ and $\mathbf{D}$.

PART III

THE TRANSITION TO MECHANICAL THEORY

CHAPTER XI

THE DYNAMICS OF THE ELECTRON

1. As in the first section it will be convenient to lead up to the bearing of the Principle of Relativity on this subject by giving an account of the earlier work.

It had been noticed as early as 1881 (by J. J. Thomson*) that the charge carried by a body has the effect of increasing the apparent inertia of the body, the electromagnetic field set up by the moving charge exercising a reaction on the charge itself.

When it was found in experimenting on the cathode rays that the carriers of the electric charge through the vacuum tube must be conceived as particles having an effective mass which was very small compared with the smallest material particle hitherto conceived, namely the atom of hydrogen, the possibility revealed itself that there might be cases in which the electromagnetic inertia of a charged particle was comparable with the whole inertia. This actually proved to be so when Kaufmann, experimenting on the β-rays emitted by radium, found them to be of similar character, save that their velocity was much greater—from about ·2 to ·6 of that of light—and that the apparent mass increased in a regular way with the velocity.

* See footnote, p. 65.

This led Abraham to take up the theoretical investigation in a systematic way, and he came to the conclusion that *the whole inertia as found by Kaufmann varied with the velocity in exactly the way in which the electromagnetic inertia should vary according to his theory.* The conclusion drawn was that *the mass of the electron is of purely electromagnetic origin**. In making his calculation, Abraham used the very natural assumption that an electron was a permanently spherical distribution of electricity.

But, in seeking to construct a theory of the constitution of matter which should explain the FitzGerald-Lorentz contraction hypothesis, Lorentz was led, as we have seen in Chapter IV, to consider two systems correlated by means of a certain algebraic transformation, and he saw clearly that his argument only stood provided that the correlation extended down to the most ultimate parts, even to the conception adopted of the electron itself. Thus he was led to the conception of a moving electron as contracting in the direction of motion to exactly the same degree as that required of a material body for the explanation of the Michelson-Morley experiment. An electron at rest being conceived as symmetrical about a central point, it thus became natural to examine the field of a small spherical distribution of electricity which, on being set in motion, is contracted to that extent, thus becoming an oblate spheroid whose axes are in the ratio

$$1 : (1 - v^2/c^2)^{\frac{1}{2}}.$$

On applying Abraham's method to this conception of the electron, Lorentz obtained a result quite different in form from that obtained by Abraham from the conception of a spherical electron.

Kaufmann therefore instituted further experiments with the object of deciding between the two conceptions of the electron, and his results seemed to favour that of Abraham. But recent experiments by Bucherer and others are at least as favourable to that of Lorentz as to that of Abraham†.

* Abraham, *Theorie der Elek.* Vol. II. p. 128 (2nd ed.).

† For details, see pp. 151–2.

In either theory it appears that we no longer have a single number or even a single function of the velocity which gives the ratio of the external force on the electron to the acceleration produced, but that in fact the acceleration is only in the direction of the force when that direction is parallel to or perpendicular to the velocity, and that the ratio of force to acceleration is different in these two cases. Now Lorentz was able to shew that this difference between *longitudinal* and *transverse* mass of the electrons was exactly what was required in the theory of the Rayleigh-Brace experiment to neutralize the aeolotropic distribution of them caused by the shrinkage of the refracting medium in the direction of its motion, and the null-result was so explained. On the other hand, if the ratio of the two masses as obtained by Abraham were adopted, a positive result would be expected to this experiment*.

In just the same way, if the conductivity of a bar of metal is explained by the existence within it of free electrons, the shortening of the bar in one direction is compensated by the change in the effective mass of the electrons; with the result, verified experimentally by Trouton and Rankine, that there is no change in the total current transmitted when the whole apparatus is rotated into different directions relative to the velocity.

It will be seen then that, step by step, Lorentz was able to give an account of all the unexpected experimental results. In the attempt to explain the FitzGerald contraction for a material body he is led to assume it for the individual electron, and hence is able to explain various other null results for matter in bulk. But as he himself points out there is no *a priori* justification for the assumption, the difficulty is in fact only thrown back another stage to a point where it is useless to analyse further†. To try to explain the contraction of the electron by further assumptions about forces would only be to take one more step in an unending sequence.

* *Theory of Electrons*, §§ 184–6, pp. 216–20.

† *Theory of Electrons*, § 182, pp. 214–5.

There are other assumptions, however, beside that of the contracting electron, made in the course of Lorentz' argument which must be considered in order that we may see properly the relation between the historic development of the electron theory and the *principle of relativity* which covers and generalizes all the assumptions and results. It will therefore be convenient at this point to give a brief account of the theory of the dynamics of the electron.

2. Aethereal momentum.

In a thorough-going electrical theory of matter, such as is contemplated by some, it is necessary to eliminate at the outset the terms 'mass,' 'force,' 'momentum,' and possibly also 'energy,' at least as far as it is associated with the idea of 'work done.' But in deriving equations to be compared with the classical mechanical theory, we soon find it convenient to re-introduce these terms.

In Lorentz' theory the first step in the passage from pure electromagnetic theory to a derived mechanics is in his well-known equation

$$\mathbf{F} = \mathbf{E} + [\mathbf{uH}]/c,$$

where $\mathbf{F}$ is called the 'ponderomotive force.' This equation can now be taken as nothing more than a definition of what is meant by this term 'force,' although, in its origin, when mechanical properties were given priority, it stood rather for a definition of electromotive intensity, and contained within itself a statement of the manner in which the force on a moving charge varied with its velocity.

But now the question as to the relation of the acceleration of a moving charged particle in a given field of force cannot be taken to be settled by the statement that '$\mathbf{F}$ is the moving force,' for the classical experiments on the deviation of the cathode rays demonstrate the failure of Newtonian dynamics in this case.

There are some important relations however involving this vector which are vital to the transition from the general electromagnetic equations to dynamics.

We shall see (pp. 158–60) that the adoption of the expression $\mathbf{E}+[\mathbf{uH}]/c$ for the 'force per unit charge' is consistent with the expression $\iiiint \frac{1}{2}(\mathbf{H}^2-\mathbf{E}^2)\,dx\,dy\,dz\,dt$ for the 'action' of the whole system which we have already seen becomes a minimum subject to the assumed equations of the field*.

It may also be shewn that the quantities $[\mathbf{EH}]/c$ and $\frac{1}{2}(\mathbf{E}^2+\mathbf{H}^2)$ may also consistently with this expression for the 'force' be adopted as expressions for the 'momentum' and 'energy per unit volume' distributed throughout space.

By the application of the fundamental equations we may shew that ρF_x, the force per unit volume in the direction of the axis of x on the electricity in the field, is

$$\rho F_x = \frac{\partial X_x}{\partial x} + \frac{\partial X_y}{\partial y} + \frac{\partial X_z}{\partial z} - \frac{\partial g_x}{\partial t} \quad \text{............(1)},$$

where $\mathbf{g}=[\mathbf{EH}]/c$,

$$X_x = \tfrac{1}{2}(E_x^2+H_x^2-E_y^2-H_y^2-E_z^2-H_z^2),$$
$$X_y = E_xE_y + H_xH_y,$$
$$X_z = E_xE_z + H_xH_z.$$

The analysis is as follows:

$$\rho\mathbf{F} = \mathbf{E}\,\mathrm{div}\,\mathbf{E} + [\rho\mathbf{u},\,\mathbf{H}]/c$$
$$= \mathbf{E}\,\mathrm{div}\,\mathbf{E} + \left[\mathrm{curl}\,\mathbf{H} - \frac{1}{c}\frac{\partial\mathbf{E}}{\partial t},\,\mathbf{H}\right]$$
$$= \mathbf{E}\,\mathrm{div}\,\mathbf{E} + [\mathrm{curl}\,\mathbf{H},\,\mathbf{H}] + \left[\mathbf{E},\,\frac{1}{c}\frac{\partial\mathbf{H}}{\partial t}\right] - \frac{1}{c}\frac{\partial}{\partial t}[\mathbf{EH}]$$
$$= \mathbf{E}\,\mathrm{div}\,\mathbf{E} + \mathbf{H}\,\mathrm{div}\,\mathbf{H} + [\mathrm{curl}\,\mathbf{H},\,\mathbf{H}] + [\mathrm{curl}\,\mathbf{E},\,\mathbf{E}] - \frac{1}{c}\frac{\partial}{\partial t}[\mathbf{EH}],$$

introducing the zero term $\mathbf{H}\,\mathrm{div}\,\mathbf{H}$ for the sake of symmetry.

Now $\mathbf{E}_x\,\mathrm{div}\,\mathbf{E}+[\mathrm{curl}\,\mathbf{E},\,\mathbf{E}]_x$

can be seen at once to be exactly

$$\frac{\partial}{\partial x}\tfrac{1}{2}(E_x^2-E_y^2-E_z^2) + \frac{\partial}{\partial y}(E_xE_y) + \frac{\partial}{\partial z}(E_xE_z).$$

Adding the corresponding expression in $\mathbf{H}$ the above result follows.

* Part II, § 8, pp. 111–2.

Note that it is only proved above that the adopted expressions are *consistent*, not that they are the only ones possible, mathematically they clearly are not, but all that we are concerned with here, as in all theory, is to obtain a consistent scheme. See Macdonald, *Electric Waves*, Ch. IV, § 23, and Schott, *Electromagnetic Radiation*, pp. 5–9. In Chapter XV (Relativity and an Objective Aether) the expressions for the flux of energy and momentum are associated with a definite state of stress and velocity in the aether, so that they become to that extent less ambiguous.

We may interpret this result as follows in order to bring it into line with the ordinary dynamical theorems of momentum.

The theorem of equality of action and reaction as between the aether and the carriers of charge is maintained if we call $-\rho\mathbf{F}$ the 'force per unit volume on the aether.'

If we take any closed volume *fixed in space* and integrate the equation (1) through it we obtain

$$\iiint -\rho F_x dV = \frac{\partial}{\partial t}\iiint g_x dV - \iint (\mathbf{X}d\mathbf{S}) \quad \text{.........(2).}$$

Thus the theorem of momentum is maintained as regards the aether if we call $\iiint \mathbf{g}\, dV$ the *total momentum* within the volume and $-\mathbf{X} = -(X_x, X_y, X_z)$ the *rate of transport of the x-component of momentum of the aether across unit area.*

In the same way we have $-(Y_x, Y_y, Y_z)$ as the rate of transport of the y-component of momentum where

$$Y_y = \tfrac{1}{2}(E_y^2 + H_y^2 - E_z^2 - H_z^2 - E_x^2 - H_x^2),$$
$$Y_x = X_y = E_x E_y + H_x H_y,$$
and
$$Y_z = E_y E_z + H_y H_z,$$

and similar expressions for the z-component.

The set of quantities

$$\begin{vmatrix} X_x, & X_y, & X_z \\ Y_x, & Y_y, & Y_z \\ Z_x, & Z_y, & Z_z \end{vmatrix}$$

form what is known in vector analysis as a *tensor**.

* Unfortunately the word '*tensor*' has been used in two senses. Heaviside uses the term '*tensor of a vector*' to mean merely its magnitude or absolute value. The sense in which the term is used here was introduced by Willard Gibbs in his *Vector Analysis*.

The tensor found above is exactly the set of quantities suggested by Maxwell as defining the 'stress' in the aethereal medium. In fact the stress across an element of area of a material medium in any direction measures the rate of transfer of momentum across that element—*provided the element moves with the matter in question.*

The elements of area used in computing the above tensor were elements at rest in space, so that *the tensor is only legitimately called the stress* if the aether is considered to be at rest. It has been customary to pass over the objection to ascribing momentum to a medium supposed to be at rest, by a reminder that the momentum is a fictitious one, having no actual significance in dynamics, or else by saying that the aether is only approximately at rest. While this may be satisfactory when we are only looking towards immediate applications, from the theoretical point of view it is not. The ultimate object in introducing the terms *momentum, stress,* etc., is to examine how far it is possible to reconcile the concept of the aether with the general principles of dynamics, and it would at least be a considerable divergence from those principles to begin by assigning momentum to a medium at rest and thinking of work done by forces without motion of the medium on which they act. We shall see below* that for the closest reconciliation, it is necessary to assign a definite velocity to the aether at all points, and then the tensor found above will no longer be capable of being called the 'stress-tensor.' It will perhaps be best to call it the 'Maxwell-tensor' for the present.

3. Aethereal energy.

The rate of work of the force $\rho\mathbf{F}$ on the charge per unit volume which is moving with velocity $\mathbf{u}$ is

$$\begin{aligned}(\rho\mathbf{F}, \mathbf{u}) &= (\rho\mathbf{u}, \mathbf{E} + [\mathbf{uH}]/c) \\ &= (\rho\mathbf{u}, \mathbf{E}) \quad \text{since} \quad (\mathbf{u}[\mathbf{uH}]) = 0 \\ &= \left(c \operatorname{curl} \mathbf{H} - \frac{\partial \mathbf{E}}{\partial t}, \mathbf{E}\right).\end{aligned}$$

* Chapter XV.

This can be transformed (see below*) into the expression

$$-\operatorname{div} c\,[\mathbf{EH}] - \frac{\partial}{\partial t}\tfrac{1}{2}(\mathbf{E}^2 + \mathbf{H}^2).$$

This being the work done on the charge by the aethereal forces, the work done on the aether by the moving charge is

$$-(\rho\mathbf{F}, \mathbf{u}) = \operatorname{div} c\,[\mathbf{EH}] + \frac{\partial}{\partial t}\tfrac{1}{2}(\mathbf{E}^2 + \mathbf{H}^2) \quad \text{.........(3)},$$

or integrating through any *fixed* closed volume

$$\frac{\partial}{\partial t}\iiint \tfrac{1}{2}(\mathbf{E}^2 + \mathbf{H}^2)\,dV = -\iiint (\rho\mathbf{F}, \mathbf{u})\,dV - \iint c^2(\mathbf{g}d\mathbf{S}) \quad \text{...(4)},$$

where $\mathbf{g} = [\mathbf{EH}]/c$ as above.

This equation is commonly interpreted as follows:

$$w = \tfrac{1}{2}(\mathbf{E}^2 + \mathbf{H}^2)$$

is called the 'energy per unit volume in the aether' and the equation therefore expresses that the rate of increase of energy within a fixed volume is equal to the rate of work done by the force on the aether diminished by an amount which can be interpreted as due to a 'flow of energy' across the surface and represented by the vector $c^2\mathbf{g}$†. This is the famous *Poynting Vector*. It will be denoted in future by $\mathbf{Q}$.

To bring this into relation with the preceding and with ordinary dynamics it should be shewn that this flow of energy can be attributed to the stress in the aether, but this question will be deferred to a later chapter‡.

4. Integration of the above equations.

If we integrate the equations (2) and (4) through the whole of space, and *assume that the field is such that*

$$\iint (\mathbf{X}d\mathbf{S}) \to 0 \quad \text{and} \quad \iint (\mathbf{g}d\mathbf{S}) \to 0,$$

* For $(\mathbf{E}\operatorname{curl}\mathbf{H} - \mathbf{H}\operatorname{curl}\mathbf{E}) = -\operatorname{div}[\mathbf{EH}].$

$$\therefore\ (\mathbf{E}\operatorname{curl}\mathbf{H}) = -\operatorname{div}[\mathbf{EH}] + (\mathbf{H}, \operatorname{curl}\mathbf{E})$$
$$= -\operatorname{div}[\mathbf{EH}] - \frac{1}{c}\left(\mathbf{H}, \frac{\partial\mathbf{H}}{\partial t}\right)$$

† See note p. 140, ll. 1–8.

‡ Chap. XV, pp. 193–8.

as the boundary is indefinitely enlarged, we have in the limit

$$\iiint \rho \mathbf{F} dV = -\frac{\partial}{\partial t}\iiint \mathbf{g}\, dV \quad \text{..............(5)},$$

and

$$\iiint (\rho \mathbf{F}, \mathbf{u})\, dV = -\frac{\partial}{\partial t}\iiint w\, dV \quad \text{..............(6)}.$$

Suppose now that the carriers of the charge have momentum of the ordinary dynamical type, distributed with density $\mathbf{g}'$ per unit volume, and that in addition to the reaction between them and the aether represented by the electromagnetic force $\mathbf{K} = \rho \mathbf{F}$ there are forces of non-electromagnetic type represented by $\mathbf{K}'$ per unit volume, so that the equation of motion for these carriers becomes

$$\mathbf{K} + \mathbf{K}' = \frac{\partial \mathbf{g}'}{\partial t},$$

which must be satisfied at all points at which there is a charge carrier. This combined with (1) gives

$$\frac{\partial}{\partial t}(g_x + g_x') = K_x' + \frac{\partial X_x}{\partial x} + \frac{\partial X_y}{\partial y} + \frac{\partial X_z}{\partial z},$$

which integrated through the whole of space gives

$$\frac{\partial G_x}{\partial t} = \iiint K_x' dV,$$

assuming as before the vanishing of the surface integral $\iint \mathbf{X} d\mathbf{S}$ over the infinite boundary, where

$$(G_x, G_y, G_z) = \mathbf{G} = \iiint (\mathbf{g} + \mathbf{g}')\, dV,$$

which may be called the 'total momentum.'

5. Case in which the non-electromagnetic forces have zero sum.

A case of particular importance is that in which the non-electromagnetic forces $\mathbf{K}'$ have their sum zero. This would be so in particular if there were no such forces. In this case we should have

$$\frac{\partial \mathbf{G}}{\partial t} = 0,$$

that is, $\mathbf{G}$ would be constant throughout the history of the system, or *the sum of the electrical and non-electrical momenta is constant if there are no forces of non-electromagnetic origin.*

6. The case of purely electromagnetic inertia.

Another important case for consideration is that in which there is assumed to be *no inertia of non-electromagnetic origin*, as suggested by Abraham and Kaufmann. In that case we have, since $\mathbf{g}' = 0$,

$$\mathbf{K} + \mathbf{K}' = 0$$

at all points at which there is a carrier for the charge.

A similar result follows in the case of the energy: putting w' for the density of non-electrical energy, we have

$$\frac{\partial w'}{\partial t} = (\mathbf{K} + \mathbf{K}', \mathbf{u}),$$

giving

$$\frac{\partial}{\partial t}(w + w') = (\mathbf{K}', \mathbf{u}) - c^2 \operatorname{div} \mathbf{g}.$$

On integrating through all space, and calling W the total energy, we have

$$\frac{\partial W}{\partial t} = \iiint (\mathbf{K}'\mathbf{u})\, dV,$$

that is *the rate of increase of the total energy is equal to the rate of working of the non-electrical forces.*

7. The application of the above results by Lorentz and Abraham to the dynamics of the electron.

The first step is the calculation of the momentum in the aether due to a single electron which is supposed to have moved always with constant velocity. Given the whole motion of an electron the fundamental equations are sufficient to determine that field, and by the application of a Lorentz transformation this field can be correlated with an electrostatic field which is completely known.

(*a*) *Abraham's electron.*

If we consider the electron as suggested by Abraham to be a spherical distribution of radius a whatever its velocity relative to the observer, the correlated field is the electrostatic field of a prolate ellipsoid of axes a and βa, where $\beta = (1 - v^2/c^2)^{-\frac{1}{2}}$, with the charge of electricity distributed as it would be if the spheroid were a conductor*.

This field being known, that of the moving sphere can be written down by reversing the transformation, and hence the momentum found. The result is

$$|\mathbf{G}| = \frac{e^2}{16\pi ac}\left\{\frac{c^2+v^2}{v^2}\log\frac{c+v}{c-v} - \frac{2c}{v}\right\},$$

the direction of $\mathbf{G}$ being that of the velocity.

The corresponding expression for the energy of the field is

$$W = \frac{e^2}{8\pi a}\left\{\frac{c}{v}\log\frac{c+v}{c-v} - 1\right\}\dagger.$$

(*b*) *Lorentz' electron.*

The electron as conceived by Lorentz is such that the correlated system is a sphere of radius a at rest with uniform charge and in this case the momentum is found to have the simple value

$$\mathbf{G} = \beta\frac{e^2\mathbf{v}}{6\pi ac^2},$$

while the energy is

$$W = \beta\frac{e^2}{6\pi a}\left(1 + \frac{3v^2}{c^2}\right).$$

The calculations are given below for this case, those for Abraham's electron being omitted as they do not concern us here‡.

Of these two conceptions the former is based on the Newtonian conception of a rigid body, whereas the latter corresponds to a body the configuration of which is permanent as that term is understood in the principle of relativity. The Abraham model is based on the conception of a fixed aether, and the electron is always spherical to an observer at rest relative to that aether. The Lorentz electron is always spherical to an observer *moving*

* Abraham, *Theorie der Elektrizität*, II. § 19, 2nd edition, 1908.

† First given by G. F. C. Searle, *Phil. Trans.* A, 187 (1896), p. 165.

‡ See Abraham, *Theorie der Elektrizität*, 2nd edition, pp. 165–9. For a full discussion see the whole of Chapter II of that work, or Lorentz, *Theory of Electrons*, especially §§ 26–28, 178–186, Notes 15, 16.

with it. In the light of the principle of relativity the latter is the simpler conception, not, as it appears to many, the more complex.

8. Calculation of the momentum of the Lorentz electron.

The field of Lorentz' electron in uniform motion is obtained from that of a uniformly charged sphere at rest by means of the transformations of the principle of relativity. It thus becomes simple to calculate the momentum G.

Let the electron be at rest in the system S'.

The momentum in question is

$$\frac{1}{c}\iiint [\mathbf{EH}]\,dV$$

integrated through the whole space. The components of this are, expressed in terms of the field in S',

$$\iiint \frac{1}{c}\left\{\beta^2 E_y' \cdot \frac{v}{c} E_y' - \beta^2 E_z'\left(-\frac{v}{c} E_z'\right)\right\}\frac{dV'}{\beta},$$

$$\iiint - E_x' \cdot \beta\frac{v}{c} E_y' \frac{dV'}{\beta},$$

$$\iiint E_x'\left(-\beta\frac{v}{c} E_z'\right)\frac{dV'}{\beta},$$

(E_x', E_y', E_z') being the electric intensity due to the spherical electron at rest, these integrals extending again over the whole of space in the system S'.

On integrating, the y and z components vanish by reason of symmetry, while the x component becomes

$$\frac{\beta v}{c^2}\iiint (E_y'^2 + E_z'^2)\,dV'.$$

Since, again by symmetry,

$$\iiint E_x'^2 dV' = \iiint E_y'^2 dV' = \iiint E_z'^2 dV' = \tfrac{2}{3} W',$$

the momentum becomes

$$\tfrac{4}{3}\frac{\beta \mathbf{v}}{c^2} W',$$

where W' is the total electrostatic energy of the electron at rest, which is easily found to be $\frac{1}{2}\cdot\frac{e^2}{4\pi a}$ or $\frac{3}{5}\cdot\frac{e^2}{4\pi a}$ according as the distribution is uniform over the surface or through the volume.

In either case therefore the momentum of the uniformly moving electron is of the form

$$\mathbf{G} = \frac{4}{3}\frac{W'}{c^2}\frac{\mathbf{v}}{(1 - v^2/c^2)^{\frac{1}{2}}}.$$

The electrical energy of the uniformly-moving electron can be found in the same way—viz.

$$\begin{aligned} W &= \tfrac{1}{2}\iiint(\mathbf{E}^2 + \mathbf{H}^2)\,dV \\ &= \tfrac{1}{2}\iiint\left\{E_x'^2 + \beta^2(E_y'^2 + E_z'^2) + \frac{\beta^2 v^2}{c^2}(E_y'^2 + E_z'^2)\right\}\frac{dV'}{\beta} \\ &= \frac{W'}{3\beta}\{1 + 2\beta^2(1 + v^2/c^2)\} \\ &= \frac{W'}{3}\frac{3 + v^2/c^2}{(1 - v^2/c^2)^{\frac{1}{2}}}. \end{aligned}$$

9. ‘Quasi-stationary’ motion.

This term has been introduced to signify a motion of an electron which is assumed to vary so slowly that the actual momentum at any instant may be taken to be equal to the momentum, calculated as above, for an electron which moves continually with a constant velocity equal to the instantaneous velocity of the actual electron at the moment in question.

The justification of this approximation lies in the fact that the effect of a change in the velocity of the particle is propagated outwards with the velocity of light, and, provided such changes are not violent, the density of the momentum diminishes very rapidly as the distance from the electron increases, so that by far the greatest part of the total momentum arises from the field in the immediate neighbourhood of the electron, and therefore depends on the motion of the electron during a very short interval of time including the instant in question*.

* Cf. Abraham, *Th. der Elek.* II. p. 198.

The hypothesis of 'quasi-stationary motion' is then that, if '$\mathbf{G}$' is the actual momentum at an instant when the velocity of the electron is $\mathbf{v}$, and '$\mathbf{G}(\mathbf{v})$' is the momentum of the same electron moving *constantly* with velocity $\mathbf{v}$,

$$\mathbf{G} = \mathbf{G}(\mathbf{v}),$$

and that if $\delta\mathbf{v}$ is the change in $\mathbf{v}$ during a time δt

$$\delta\mathbf{G} = \delta\mathbf{G}(\mathbf{v}).$$

Remembering that $\mathbf{G}(\mathbf{v})$ is a vector in the direction of $\mathbf{v}$ let $(\delta v)_l$ and $(\delta v)_t$ be the longitudinal and transverse components of $\delta\mathbf{v}$ in time δt.

Then $\delta\mathbf{G}(\mathbf{v})$ is the resultant of

$$\frac{\partial|\mathbf{G}(\mathbf{v})|}{\partial|\mathbf{v}|}(\delta v)_l \text{ longitudinally,}$$

and $$\frac{|\mathbf{G}(\mathbf{v})|(\delta v)_t}{|\mathbf{v}|} \text{ transversely*.}$$

Thus if f_l and f_t are the longitudinal and transverse accelerations, the rate of change of G is compounded of $m_l f_l$ longitudinally and $m_t f_t$ transversely, where

$$m_l = \frac{\partial|\mathbf{G}(\mathbf{v})|}{\partial|\mathbf{v}|},$$

and $$m_t = \frac{\mathbf{G}(\mathbf{v})}{|\mathbf{v}|}.$$

10. The hypothesis of electromagnetic inertia only.

As was pointed out above (p. 144), *if the carrier of the charge has no intrinsic inertia the electromagnetic force* $\mathbf{K}$ *at any point must be neutralized by the non-electromagnetic force* $\mathbf{K}'$.

In considering the electron as a whole we are bound to think of some forces as holding it together. It is these forces together with any other non-electromagnetic forces of external origin acting on the electron that are represented by $\mathbf{K}'$.

Since $\mathbf{K} + \mathbf{K}' = 0$ at every point of the electron

$$\iiint(\mathbf{K} + \mathbf{K}')\,dV = 0,$$

* The angle turned through by the direction of the velocity is $(\delta v)_t/|\mathbf{v}|$, and $(\delta v)_l = \delta|\mathbf{v}|$.

where the region of integration is the whole region occupied by the electron.

If we now assume that *the forces holding the electron together have zero sum*, $\iiint \mathbf{K}'\,dV$ gives simply the sum of the externally impressed forces on the electron.

Thus we have finally, representing this sum by $\mathbf{P}$,

$$\mathbf{P} = -\iiint \mathbf{K}\,dV$$

$$= \frac{d\mathbf{G}}{dt}$$

$$= m_l f_l + m_t f_t,$$

this being a vector addition.

The two functions of the velocity, m_l, m_t, are known as the *longitudinal* and *transverse masses*.

For the Lorentz electron

$$m_l = \frac{e^2}{6\pi a c^2}\beta^3,$$

$$m_t = \frac{e^2}{6\pi a c^2}\beta;$$

since these quantities are unequal (except for $v = 0$, when $\beta = 1$) the vector $\mathbf{P}$ is not in the direction of the resultant acceleration unless either f_l or f_t is zero.

11. Abraham's objection to the Lorentz electron.

The rate at which the force $\mathbf{P}$ works is, putting $v = |\mathbf{v}|$,

$$(\mathbf{P}\mathbf{v}) = m_l f_l v$$

$$= \frac{e^2}{6\pi a c^2} \frac{v\dfrac{dv}{dt}}{\left(1 - \dfrac{v^2}{c^2}\right)^{\frac{3}{2}}}$$

$$= \frac{d}{dt}\left(\frac{e^2}{6\pi a}\beta\right).$$

The expression for the total energy of the electromagnetic field is given above as

$$\frac{e^2}{6\pi a}\beta\left(1+\frac{v^2}{3c^2}\right);$$

so that the work of the *external* forces $(\mathbf{Pv})$ is not equal to the rate of increase of the *total* energy. Abraham has therefore raised the objection that the electron of Lorentz does not satisfy the principle of the conservation of energy; or, more precisely, that the electron cannot be called a *purely electromagnetic* entity, since the work of the forces which hold it together appears in the equation of energy.

In fact if $\mathbf{K}_1'$ is the non-electromagnetic force which represents the effect of the kinematic constraint on the charge of the electron, the assumption

$$\iiint \mathbf{K}_1' dV = 0$$

does not involve the condition

$$\iiint (\mathbf{K}_1'\mathbf{w})\, dV = 0,$$

where $\mathbf{w}$ is the velocity of a point of the electron; for, with a contracting configuration, the velocity $\mathbf{w}$ will vary from point to point.

In Abraham's model of a *rigid* electron on the other hand, $\mathbf{w}$ is conceived as constant throughout, so that the work of the constraining forces is zero, and we do find in that case that the rate of increase of the electromagnetic energy is equal to $(\mathbf{Pv})$.

It is however not strictly correct to speak of either model as 'purely electromagnetic,' inasmuch as the ordinary laws of electromagnetism are not by themselves sufficient for the preservation of the geometrical characteristics of either.

Only it is to be remembered that in dealing with the dynamics of the Lorentz electron, the total energy is not simply the electromagnetic energy as given above, but the sum of this and the potential energy of the constraining forces, and this sum is exactly $e^2\beta/6\pi a$, taking the zero of potential energy to be the spherical configuration corresponding to $v=0$.

12. The results of experiments on the apparent mass of the negative electrons.

For the express purpose of discriminating between the results of Lorentz and Abraham for the *transverse mass* of an electron, Kaufmann repeated his experiments on the deviation of the β-rays by electric and magnetic fields*. The conclusion to which he came was that Abraham's expression more nearly represented the facts observed. But the great importance attaching to the answer to the question led to other investigations, and Bucherer, Ratnowsky, and Hupka declared in favour of Lorentz' expression. For comparison's sake it may be worth while to give the figures as determined†. In all cases the charge is measured in electromagnetic units.

Kaufmann's results.

If m is the effective *transverse* mass of the electron for the velocity v and m_0 the value for $v = 0$ so that

$$m = m_0 f(v),$$

where $f(v)$ is one or other of the functions obtained by Lorentz and Abraham, the experiment allows of the determination of v and e/m, where e is the charge.

Kaufmann found that by taking different values for m_0 with the two different forms of $f(v)$ he could make the observed variation in e/m agree equally well in both cases with the calculated.

The values for e/m_0 were respectively

on Lorentz' theory ... ... $1{\cdot}660 \times 10^7$,

on Abraham's theory ... ... $1{\cdot}823 \times 10^7$.

For the more slowly moving cathode-rays, where the variation in m is negligible there is no question as to any particular theory and the corresponding number is according to Kaufmann $1{\cdot}878 \times 10^7$.

* Kaufmann, *Ann. d. Phys.* 19 (1906), p. 487; 20 (1906), p. 639.

† For a general account of these experiments up to 1910 and of criticisms raised, see J. Laub, "Die experimentelle Grundlagen der Relativitätsprinzip," *Jahrbuch der Radioaktivität und Elektronik*, 7 (1910), p. 405.

Kaufmann's conclusion was that Abraham's theory was in fair accord with the known value of e/m_0 while Lorentz' was not.

Bucherer's results (*Ann. d. Phys.* 28 (1909), p. 513).

By a modified method, Bucherer obtained the following figures:

v/c	e/m_0 (on Lorentz' theory)	e/m_0 (on Abraham's theory)
·3787	$1{\cdot}701 \times 10^7$	$1{\cdot}675 \times 10^7$
·4281	·699	·663
·5154	·700	·645
·6780	·701	·580

His conclusion was that since Lorentz' theory gives the more constant value of e/m_0 it is supported by the experiments.

Bestelmeyer's results (*Ann. d. Phys.* 22 (1907), p. 429).

v/c	e/m (observed)	e/m (Lorentz)	e/m (Abraham)
·1951	$1{\cdot}697 \times 10^7$	$1{\cdot}694 \times 10^7$	$1{\cdot}700 \times 10^7$
·2469	·678	·678	·679
·3222	·643	·647	·640

Here the values of e/m in the last two columns are calculated from the value

$$\frac{e}{m_0} = 1{\cdot}72 \times 10^7,$$

which Bestelmeyer determined from experiments on the slow cathode-rays. It will be seen that in each case the observed value of e/m lies between the two calculated values which are very near to one another.

But if $\frac{e}{m_0}$ is equal to $1{\cdot}72 \times 10^7$, and not to $1{\cdot}878 \times 10^7$ as stated above, Kaufmann's experiments seem to favour Lorentz' result rather than Abraham's, and in fact all later experiments seem to point to this value being the more accurate.

Hupka finds from a large number of observations that Lorentz' theory seems to give a more constant value of e/m_0 than does Abraham's (*Ann. d. Phys.* 31 (1910), p. 169).

Ratnowsky (*Comptes Rendus*, 1910) also pronounces in favour of Lorentz' theory.

The value of all the preceding results must however be considered in the light of more recent experiments which have benefited by the experience of the earlier ones.

Wolz following up Bucherer's method finds that from $v/c = \cdot 5$ to $v/c = \cdot 7$, Lorentz' theory gives a very constant value of e/m_0, namely $1{\cdot}767 \times 10^7$ (*Ann. d. Phys.* 30 (1909), p. 273).

The following values of e/m_0 have been obtained by the writers named:

$1{\cdot}773 \times 10^7$	Classen, *Verh. d. Deut. Phys. Ges.* 10 (1908), p. 700
$1{\cdot}769 \times 10^7$	J. Malassez, *Ann. de Chim. et de Phys.* 23 (1911), p. 231
$1{\cdot}766 \times 10^7$	A. Bestelmeyer, *Ann. der Phys.* 35 (1911), p. 909
$1{\cdot}766 \times 10^7$*	J. Alberti, *Ann. der Phys.* 39 (1912), p. 1133

The most recent experiments of all, those of G. Neumann†, are very decisive in favour of the predictions of the principle of relativity. For β-rays of velocities varying from $\cdot 4c$ to $\cdot 8c$ the values of $\frac{e}{m_0}$ calculated according to Abraham's theory vary between $1{\cdot}55 \times 10^7$ and $1{\cdot}74 \times 10^7$; while according to Lorentz formula the results of 26 experiments lie between 1·75 and 1·8 and all but four of these between 1·75 and 1·78, with a mean of $1{\cdot}765 \times 10^7$. This agrees so closely with the values of e/m_0 just above, that there seems little room left for doubt now that the principle of relativity is in accord with the facts of experiment.

* Given by Alberti as $1{\cdot}756 \times 10^7$, but corrected by Schaefer in the paper cited below.

† The results are recorded by C. Schaefer, *Phys. Zeits.* 14 (1913), p. 1177.

CHAPTER XII

RELATIVITY AND DYNAMICAL THEORY

1. The necessity for a revision of dynamical theory.

The decision between the two theories of the electron is not so important as the result, about which there is no doubt, namely that it is *impossible to deal with the inertia of the electron as if it were represented by a simple Newtonian mass.*

We are not necessarily bound to admit the conclusion of Abraham and Kaufmann in 1902 that "the mass of the electron is of *purely* electromagnetic origin"—this might be only approximately true, and we have seen too that *pure* electromagnetic theory is not sufficient to determine the motion of an electron. But we have at least a glimpse of a region in which Newtonian dynamics is not fundamental or universal. Further, with the growth of our knowledge of the complexity of the material atom, it seems possible that the sum of the variable parts of the masses of the electrons within an atom is comparable with the whole mass of the atom, and indeed that the same may be true of an extended material body. Without admitting the ambitious programme of the *completely* electromagnetic constitution of matter, it must at once be recognized that the mass of a material body becomes a statistical quantity depending partly on the motions of the multitude of electrons within it, and that only the fact of the smallness of the relative velocities of all bodies of which ordinary dynamics treat, compared with that of light, has made it possible to think for so long of the constant mass-ratio of two bodies as a universal property of matter.

If this is granted, we are bound to seek for the relation of dynamical theory to these new facts, and to regulate suggestions as to the nature of the electron and of the constitution of matter not only by select *ad hoc* experiments but also by their bearing on those concepts and theories which have hitherto been given a fundamental place.

2. There are two interacting processes at work in theoretical science. There is the continual endeavour to form models which, working according to the commonly adopted laws of well-established phenomena, shall imitate or represent, more or less closely, phenomena of which the *modus operandi* is not clearly realized—for example, Lord Kelvin's gyrostatic aether, the elastic-solid theory of light, the earlier emission theory of light. The conceptions of the electron used by Lorentz and Abraham belong to this aspect of science, and the conception of matter as constituted of electrons, though incomplete at present, is an attempt at such a model in respect of matter.

There is on the other hand the attempt to disentangle general principles of the widest possible application, not fully descriptive of each particular set of phenomena, but common to them all, and independent of the special mechanism which is characteristic and particular. Newton's Laws of Motion may be instanced, the principle of least action, the principle of the conservation of energy. With these we may class the Principle of Relativity, that is, the general hypothesis, suggested by experience, *that whatever be the nature of the aethereal medium we are unable by any conceivable experiment to obtain an estimate of the velocities of bodies relative to it.*

3. We have seen above that the models of the electron suggested by Lorentz and Abraham would, if valid, imply a revision of the Newtonian conception of mass and therefore of the whole system of dynamics.

The question then arises, what has the general Principle of Relativity, if adopted, to say as to such a necessity, independently of any particular model of a material system? Is it possible that some of the conclusions which have been drawn by Lorentz from his particular model may be consequences of this general principle alone, and that the reason they arise out of the argument of Lorentz is that that model is itself subject to the general principle?

4. The four-dimensional analysis introduced by Minkowski not only introduces a greater symmetry into the discussion of the relativity of electrodynamic phenomena. It gives us also a new point of view from which to regard mechanical quantities and enables us to go some way in finding what modifications are necessary to the usual statements of mechanical theory in order that they may be included within the scope of the principle of relativity.

Such modifications cannot pretend to be proved *necessary* throughout the domain of mechanics any more than the hypotheses of Lorentz are the only ones possible, but in the positive results of the experiments on the apparent mass of the negative electron, and in all attempts at a constitutive theory of the negative results of the Michelson-Morley and Rayleigh-Brace experiments, it is impossible to maintain the old conceptions in their entirety and it is therefore a step forward to employ a general principle which covers the results of all those experiments, and a method which can be used in criticizing suggested adjustments.

Further, the symmetry which can be introduced into mechanics by the adoption of this point of view is no less remarkable than that which we have observed above in electromagnetic theory, and the duality noticeable in the history of dynamics in the contrast between Huygens and Galilei, Euler and D'Alembert, the school of 'energetics' and the school of 'forces,' is turned into a striking unity.

We notice for instance that the fundamental difference between Huygens and Galilei was that the former concerned himself with 'space rate of change' of what we now call 'energy,' while the latter dealt with 'time rate of change' of 'momentum.' Now in the four-dimensional world of Minkowski, space and time combine into a single concept in which displacement in space and displacement in time are exactly on the same footing, and the suggestion arises that a force and its rate of working may also combine into a single concept in which the distinction between the two becomes lost, emerging only when we again separate the notions of space and time. In fact just as three-dimensional kinematics becomes four-dimensional geometry of rest, so three-dimensional dynamics may become four-dimensional statics.

5. Illustration—the Principle of Least Action.

An example may at once be given from the region of electrodynamics. It is known that the electrostatic field due to a given distribution of charge is uniquely determined as that which gives the potential energy, represented by the space integral $\frac{1}{2}\iiint \mathbf{E}^2 dV$, a stationary value, subject to $\operatorname{div} \mathbf{E} = \rho$, and that if the distribution of the charges is varied, the variation of the integral gives the virtual work of the forces acting on the charges. We may compare with this the relation of the electrodynamic equations to the principle of least action which is contained in the following theorem.

If $\mathfrak{F}$ *is the* 6-*vector* $(\mathbf{H}, -i\mathbf{E})$, *subject to the equation*

$$\operatorname{lor} \mathfrak{F} = \mathfrak{s},$$

$\mathfrak{s}$ *being the* 4-*vector* $\rho(\mathbf{v}/c, i)$, *then if* $\mathfrak{F}$ *is determined so that the space-time integral* $\frac{1}{2}\iiiint \mathfrak{F}^2 dx\,dy\,dz\,dt$ *is stationary for small variations of* $\mathfrak{F}$ *when* $\mathfrak{s}$ *is kept constant,* $\mathfrak{F}$ *will satisfy the equation*

$$\operatorname{lor} \mathfrak{F}_1 = 0\,;$$

and further, if $\mathfrak{s}$ *be varied, the variation of the integral will give the virtual work of the forces acting on the charges.*

The invariance of all the expressions occurring in the enunciations of the above theorems illustrate the criterion which is to be adopted in what follows; but before proceeding with the exposition of it, it may be useful to give a proof of the second part of the theorem just stated. The first part has been proved above (pp. 111–2).

6. Relation of the fundamental equations to Hamilton's Principle.

Taking the invariant quantity

$$A = \frac{1}{2}\iiiint \mathfrak{F}^2 dx\, dy\, dz\, du,$$

we proceed to consider its variation when we suppose that the element of charge which in the actual state of the system is at the point $(x, y\ z)$ at time t is displaced so as to be at the point $(x + \delta x, y + \delta y, z + \delta z)$ at time $t + \delta t$ supposing always that in the varied system the equations

$$\text{lor}\,(\mathfrak{F} + \delta\mathfrak{F}) = \mathfrak{s} + \delta\mathfrak{s}$$

and $$\text{lor}\,(\mathfrak{F}_1 + \delta\mathfrak{F}_1) = 0$$

are satisfied.

We have, since lor $\mathfrak{F}_1 = 0$,

$$\mathfrak{F} = \text{curl}\ \mathfrak{a},$$

and, since lor $(\mathfrak{F}_1 + \delta\mathfrak{F}_1) = 0$,

$$\delta\mathfrak{F} = \text{curl}\ \delta\mathfrak{a},$$

and $$\text{lor curl}\ \delta\mathfrak{a} = \delta\mathfrak{s},$$

$\delta\mathfrak{a}$ being a variation in the electrodynamic potential $\mathfrak{a}$.

Thus

$$\delta A = \iiiint (\mathfrak{F}\,\delta\mathfrak{F})\, dx\, dy\, dz\, du$$

$$= \iiiint (\text{curl}\ \mathfrak{a}\ \text{curl}\ \delta\mathfrak{a})\, dx\, dy\, dz\, du.$$

Applying the theorem of p. 102, § 18,

$$\delta A = \iiiint \{\text{div}\,[\mathfrak{a}, \text{curl}\ \delta\mathfrak{a}] - (\mathfrak{a}\ \text{lor curl}\ \delta\mathfrak{a})\}\, dx\, dy\, dz\, du$$

$$= -\iiiint (\mathfrak{a}\,\delta\mathfrak{s})\, dx\, dy\, dz\, du + I,$$

where I is a triple integral over the infinite boundary, which will be assumed to vanish.

7. We have now to express the variation in the current 4-vector $\mathfrak{s}$ in terms of the geometrical displacement. Since the relation must be of invariant form, it will be sufficient to consider a single component.

The simplest method is perhaps to revert for a moment to three dimensions.

Considering the *simultaneous* displacements $(\delta x - v_x \delta t, \delta y - v_y \delta t, \delta z - v_z \delta t)$ the variation in the density is clearly given by

$$\delta\rho + \frac{\partial}{\partial x}\{\rho(\delta x - v_x \delta t)\} + \frac{\partial}{\partial y}\{\rho(\delta y - v_y \delta t)\} + \frac{\partial}{\partial z}\{\rho.(\delta z - v_z \delta t)\} = 0,$$

which can be written on multiplying by ic

$$\begin{aligned}\delta s_u &= \operatorname{div}(\mathfrak{s}\,\delta u) - \operatorname{div}(s_u \delta\mathfrak{r}) \\ &= \frac{\partial}{\partial x}(s_x \delta u - s_u \delta x) + \frac{\partial}{\partial y}(s_y \delta u - s_u \delta y) + \frac{\partial}{\partial z}(s_z \delta u - s_u \delta z) \\ &= \operatorname{lor}_u [\delta\mathfrak{r}, \mathfrak{s}].\end{aligned}$$

Thus we have $\delta\mathfrak{s} = \operatorname{lor}[\delta\mathfrak{r}, \mathfrak{s}]$.

8. Applying this to the expression obtained above, we have

$$\delta A = -\iiiint (\mathfrak{a} \,.\, \operatorname{lor}[\delta\mathfrak{r}, \mathfrak{s}])\, dx\,dy\,dz\,du,$$

which on applying again the theorem (p. 102) and assuming again that the integrated part vanishes becomes

$$\begin{aligned}\delta A &= \iiiint ([\delta\mathfrak{r}, \mathfrak{s}], \operatorname{curl}\mathfrak{a})\, dx\,dy\,dz\,du \\ &= \iiiint ([\delta\mathfrak{r}, \mathfrak{s}], \mathfrak{F})\, dx\,dy\,dz\,du \\ &= \iiiint (\delta\mathfrak{r}, [\mathfrak{s}, \mathfrak{F}])\, dx\,dy\,dz\,du\, ^{*} \\ &= \iiiint (\delta\mathfrak{r}, \mathfrak{k})\, dx\,dy\,dz\,du \quad \ldots\ldots\ldots\ldots\ldots\ldots\ldots(1),\end{aligned}$$

where

$$\begin{aligned}\mathfrak{k} &= [\mathfrak{s}, \mathfrak{F}] \\ &= (s_y F_{xy} + s_z F_{xz} + s_u F_{xu}, \ldots, \ldots, \ldots) \\ &= \{\rho(\mathbf{E} + [\mathbf{uH}]/c),\ -\rho(\mathbf{Eu})\}^{\dagger},\end{aligned}$$

* $(\mathfrak{a}, [\mathfrak{b}, \mathfrak{C}])$ is easily seen from the definition to be identically equal to $([\mathfrak{a}, \mathfrak{b}], \mathfrak{C})$.

† That $\mathfrak{k}$ is a 4-vector was first noticed by Poincaré, *Rend. del Circ. mat. di Pal.* 21 (1906), p. 169.

where the notation in the last expression is that of *three-dimension* vector analysis.

9. In dynamical theory, if a system has kinetic energy T and internal potential energy V, the theorem usually known as 'Hamilton's Principle' is that the variation of the action $\delta \int (T - V)\, dt$ is equal to $-\int \delta U dt$, where δU is the virtual work of the external forces acting on the system in producing the variation from the actual position at time t to the varied position at the same instant.

Thus if $\mathbf{F}$ is the force at the point (x, y, z) of the system

$$\begin{aligned} \delta U &= F_x (\delta x - v_x \delta t) + F_y (\delta y - v_y \delta t) + F_z (\delta z - v_z \delta t) \\ &= F_x \delta x + F_y \delta y + F_z \delta z - (\mathbf{F}\mathbf{v})\, \delta t. \end{aligned}$$

Comparing this with the above equation (1) we see that *we are able to interpret that equation as being in accord with this principle of Hamilton if we call*

$$\begin{aligned} A/ic &= \iiiint \tfrac{1}{2} (\mathbf{H}^2 - \mathbf{E}^2)\, dx dy dz dt \\ &= \iiiint \tfrac{1}{2} \mathfrak{F}^2 dx dy dz dt \end{aligned}$$

the 'action' of the system, and

$$-\mathbf{F} = -(\mathbf{E} + [\mathbf{uH}]/c)$$

the 'force exerted upon the aether' per unit of charge by the bodies which carry the charge; or conversely

$$\mathbf{F} = (\mathbf{E} + [\mathbf{uH}]/c)$$

may be called the force exerted per unit of charge by aethereal action on the bodies which carry them. As in the discussion of aethereal momentum and energy (Chap. XI) the electric charge is here thought of simply as the seat of communication and transfer between the aether and the matter carrying the charge.

10. The Action an Invariant.

Following out the conception used in Chapter XI of an equilibrium between the electrical and non-electrical forces, we may put the complete conditions determining the motion of a system into the following form.

If A is the electromagnetic action as defined above and A_1 is the action of the mechanical forces, then the motion of the system is such that $(A + A_1)$ has a stationary value; that is

$$\delta (A + A_1) = 0,$$

subject to those kinematic conditions which are assumed to be permanently satisfied by the electron.

If the motion so determined is to be consistent with the principle of relativity, this statement is to have an invariant significance for the whole set of possible frames of reference. We have already seen that A is an invariant, so that the natural hypothesis to make, certainly a sufficient hypothesis, is *that A_1 is also an invariant.*

This is the simplest and most comprehensive means of extending the Principle of Relativity to cover general dynamics, but in order to obtain some of the consequences with as little calculation as possible we shall in the following chapters, instead of starting from the action, shew how the customary dynamical equations may be directly modified to meet the requirement of relativity. To do this we require a general criterion which we can apply. Such a criterion has already been used to suggest a modification of the electrodynamic equations of moving bodies. It will be as well to restate it specifically at this point.

11. The general criterion of relativity.

In order to construct a system of mechanics which is consonant with the principle of relativity, we have to ensure that all equations which are used preserve an invariant form under the transformations with which we are dealing. As in the ordinary equations of mathematical physics, we are restricted to equations connecting vector quantities and scalar quantities, so now for a similar reason *we are required to adopt equations connecting scalar quantities or 4-vectors or 6-vectors, and all the terms of any given relation must be of the same nature.*

In other words, just as in the older physics we are restricted, from the fact of the objectivity of the phenomena, to equations whose form is unaltered when we turn the axes about in any way, so in the physics subject to the principle of relativity we are restricted to equations whose form is unaltered when the various quantities involved are subjected to the transformations corresponding to any Minkowski change of coordinates. There is in fact a very close relationship between the conception of the *objectivity* of phenomena, and the *principle of relativity**.

We have been accustomed to associate objective existence in space with Euclidean geometry and Newtonian kinematics. But if we adopt the principle of relativity, *objective existence is inseparable from what we have called Einstein kinematics.* We may not conceive of a body as having physical reality unless the velocities of a given point as seen by two different observers are related to one another and to the relative velocity of the observers by the Einstein addition formulae.

Thus the critics of the Principle of Relativity are justified in saying that it does not admit of an 'objective fixed aether,' but it cannot be said that it denies the existence of an objective aether of any kind, until it is shewn that a medium cannot be conceived which renders account of electromagnetic phenomena and at the same time has a motion which is consistent with the kinematics of the principle. This point will be discussed further in a later chapter, where it will be seen that such a medium is not inconceivable.

* See below, Relativity and an Objective Aether, Chap. XV.

CHAPTER XIII

THE DYNAMICS OF A PARTICLE

1. The equations of motion in terms of 4-vectors.

We are now able to reinterpret the preliminary discussion (Chap. VI, pp. 64–70) of the motion of a material point, and to extend it to any system moving as a whole with any given velocity.

The criterion of relativity (p. 162) demands that in generalizing the Newtonian equation

$$\frac{d\mathbf{g}}{dt} = \mathbf{k},$$

where $\mathbf{g}$ is momentum and $\mathbf{k}$ is force, we must equate two 4-vectors instead of two ordinary three-dimension vectors.

Thus the simplest way of effecting this is to make the hypothesis that the momentum $\mathbf{g}$ and the energy w of a material point, or of a system whose motion is defined by a single velocity of translation, are such that $(\mathbf{g}, iw/c)$ constitutes a 4-vector $\mathfrak{g}$ which we call the 'extended momentum.'

Considering the same system with its state of motion slightly changed in any way, we have also

$$\mathfrak{g} + \delta\mathfrak{g} = \{\mathbf{g} + \delta\mathbf{g},\, i(w + \delta w)/c\} \text{ is a 4-vector}$$

and therefore

$$\delta\mathfrak{g} = (\delta\mathbf{g},\, i\delta w/c) \text{ is also a 4-vector.}$$

But further, as on p. 97,

$$(\delta x,\, \delta y,\, \delta z,\, ic\,\delta t) \text{ is a 4-vector,}$$

and therefore if $$c\delta t_0 = \{c^2\delta t^2 - \delta x^2 - \delta y^2 - \delta z^2\}^{\frac{1}{2}}$$
$$= c\delta t\,(1 - v^2/c^2)^{\frac{1}{2}},$$

δt_0 *is an invariant.* Hence on division by this invariant

$$\mathfrak{v} = \kappa\,(\mathbf{v}, ic)$$

and

$$\frac{d\mathfrak{g}}{dt_0} = \kappa\left(\frac{d\mathbf{g}}{dt}, ic\frac{dw}{dt}\right)$$

are 4-vectors.

If $\mathfrak{k}$ is a 4-vector, and we write

$$\frac{d\mathfrak{g}}{dt_0} = \mathfrak{k},$$

then we have a purely relative equation.

We will call $\mathfrak{k}$ the 'extended force vector' or more briefly the 'force 4-vector.'

2. Momentum and energy in terms of velocity.

So far nothing has been said about the relation of the 'momentum' to the 'velocity' of the system.

It is a natural extension of the Newtonian ideas to suppose that for a system at rest the momentum is zero. In the light of what has been said earlier about electromagnetic momentum, this is clearly not a hypothesis of universal application, inasmuch as the electromagnetic field which determines the momentum depends on the whole past history of the system, and it is quite easy to conceive of the nucleus of the system being at rest while there is a definite amount of momentum in the field. But at the present moment we are not laying down a general statement about all systems, but merely seeking to modify, to meet the requirement of relativity, the Newtonian conception of the momentum of a self-contained system.

Thus if a frame of reference is chosen in which the system is at rest, we have, if w_0 is the energy in that frame,

$$\mathfrak{g}_0 = (0, 0, 0, iw_0/c),$$
$$\mathfrak{v}_0 = (0, 0, 0, ic).$$

These two 4-vectors are in the same direction, and will be so in all frames of reference, their ratio being w_0/c^2.

Thus $$\mathfrak{g} = w_0 \mathfrak{b}/c^2,$$
w_0 being an invariant, or

$$\mathbf{g} = \frac{w_0 \mathbf{v}}{c^2 (1 - v^2/c^2)^{\frac{1}{2}}},$$

and $$w = \frac{w_0}{(1 - v^2/c^2)^{\frac{1}{2}}}.$$

Since w_0 is the energy of the system when considered to be at rest, it may be called the 'internal energy' of the system.

3. Inertia.

Suppose now that we confine our consideration to such systems as are completely determined by a knowledge of the velocity and internal energy, and that we consider the consequence of changing either the velocity or the internal energy.

(*a*) *Let the velocity be increased without any internal change**, that is, let the configuration of the system to an observer moving with it be unaltered.

Then the two states of the system are connected by a Minkowski transformation and w_0 is constant.

A particular case of this is the point charge discussed above (pp. 67 ff.), where any question of internal constitution is ignored. The above hypothesis of no internal change is tacitly assumed. The case of the Lorentz electron is another instance†.

(*b*) *Let the change in the system be one of internal energy without change in velocity.*

If the internal energy is changed from w_0 to $w_0 + \delta w_0$, we have as under (*a*)

$$\mathfrak{g} = w_0 \mathfrak{b}/c^2 = \frac{\kappa w_0}{c^2}(\mathbf{v}, ic) = (\mathbf{g}, iw/c),$$

and

$$\delta \mathfrak{g} = \frac{(w_0 + \delta w_0)\mathfrak{b}}{c^2} = \frac{(w_0 + \delta w_0)\,\kappa\,(\mathbf{v}, ic)}{c^2} = \left(\mathbf{g} + \delta \mathbf{g}, \frac{i(w + \delta w)}{c}\right),$$

* Such a change has been called *adiabatic*, there being no change in the internal energy, but of course more is assumed than the constancy of this one quantity.

† In this case w is not simply the electromagnetic energy, but that together with the energy of the forces maintaining the form of the system.

giving $$\frac{d\mathfrak{g}}{dt}=\frac{1}{c^2}\frac{d}{dt}\{\kappa w_0(\mathbf{v}, ic)\}=\left(\frac{d\mathbf{g}}{dt}, \frac{i}{c}\frac{dw}{dt}\right),$$

or $$\frac{d\mathbf{g}}{dt}=\frac{\kappa\mathbf{v}}{c^2}\frac{dw_0}{dt}=\frac{\mathbf{v}}{c^2}\frac{dw_0}{dt_0},$$

$$\frac{dw}{dt}=\kappa\frac{dw_0}{dt}=\frac{dw_0}{dt_0}.$$

In this case then if, as is customary, we call the rate of increase of momentum the 'force,' we see that *a force in the direction of the velocity is necessary to maintain the constant velocity of a system whose internal energy is increasing.*

(*c*) If both these types of change occur simultaneously, we have the combined result

$$\frac{d\mathbf{g}}{dt}=\frac{d}{dt}\left(\frac{w_0\mathbf{v}}{c^2(1-v^2/c^2)^{\frac{1}{2}}}\right)+\frac{\mathbf{v}}{c^2}\frac{dw_0}{dt_0}$$

$$=\frac{w_0\mathbf{f}}{c^2(1-v^2/c^2)^{\frac{1}{2}}}+\frac{\mathbf{v}}{c^2}\left\{\frac{dw_0}{dt_0}+\frac{w_0(\mathbf{vf})}{(1-v^2/c^2)^{\frac{1}{2}}}\right\},$$

giving a sum of two components in the direction of the resultant acceleration and velocity respectively; while

$$\frac{dw}{dt}=\frac{dw_0}{dt_0}+\frac{w_0}{c^2}\frac{(\mathbf{vf})}{(1-v^2/c^2)^{\frac{3}{2}}}=\frac{dw_0}{dt_0}+\left(\mathbf{v}\frac{d\mathbf{g}}{dt}\right).$$

4. The condition of constant internal energy.

Notice that with hypothesis (*a*), introducing $\mathfrak{k}$ the force 4-vector,

$$(\mathfrak{v}\mathfrak{k})=\{\kappa(\mathbf{v}, ic), \mathfrak{k}\}=\frac{\kappa w_0}{c^2}(\mathbf{v}, ic)\frac{d}{dt_0}\{\kappa(\mathbf{v}, ic)\}$$

$$=\frac{w_0}{c^2}\frac{d}{dt_0}\{\tfrac{1}{2}\kappa^2(\mathbf{v}, ic)^2\}$$

$$=w_0\frac{d}{dt_0}(-\tfrac{1}{2})$$

$$=0,$$

that is $$k_x v_x+k_y v_y+k_z v_z+k_u v_u=0,$$

or $$k_u=\frac{i}{c}(k_x v_x+k_y v_y+k_z v_z).$$

We express this by saying that the *force 4-vector is orthogonal to the velocity 4-vector.* This is the necessary and sufficient condition that the internal energy of the system shall be unchanged.

In fact, in the general case

$$(\mathfrak{v}\mathfrak{k}) = \left(\mathfrak{v},\ \frac{d}{dt_0}\frac{w_0\mathfrak{v}}{c^2}\right)$$

$$= \frac{\mathfrak{v}^2}{c^2}\frac{dw_0}{dt_0} + \frac{w_0}{c^2}\left(\mathfrak{v}\frac{d\mathfrak{v}}{dt_0}\right)$$

$$= -\frac{dw_0}{dt_0},$$

since
$$\mathfrak{v}^2 = -c^2.$$

Hence if
$$(\mathfrak{v}\mathfrak{k}) = 0,$$

$$\frac{dw_0}{dt_0} = 0.$$

5. The general significance of Lorentz' electron.

We now see that if, and only if, we treat the electron as a small body whose internal energy is unaltered during a process of acceleration, we may equate the rate of increase of its extended momentum, $\kappa\frac{d\mathfrak{g}}{dt}$, to the 4-vector $\mathfrak{k}q$ (cf. Assumption (*a*), p. 67). For $\mathfrak{k}$ is orthogonal to the extended velocity 4-vector and only in this case is the 4-vector $\kappa\frac{d\mathfrak{g}}{dt}$ also orthogonal to it.

Thus we obtain the equations

$$q\left(\mathbf{E} + [\mathbf{vB}]/c\right) = \frac{d\mathbf{g}}{dt}$$

$$= \frac{d}{dt}\left(\frac{\kappa w_0\mathbf{v}}{c^2}\right),$$

with the energy equation deducible therefrom

$$q\,(\mathbf{Ev}) = \frac{d}{dt}(\kappa w_0),$$

as the general form of (4) and (5) on p. 69.

These equations are all contained in the single equation

$$\mathfrak{k}q = \frac{d\mathfrak{g}}{dt_0},$$

$\mathfrak{k}$ *representing the 'extended force' as above,* $\mathfrak{g}$ *the 4-vector* $(\mathbf{g}, iw/c)$ *which we might call the 'extended momentum,' and* δt_0 *denoting the invariant element* $\delta t/\beta$ *which Minkowski calls the element of 'proper time' (Eigenzeit) of the moving system.*

6. The modification of the Newtonian Laws of Motion.

More generally we see that we have equations of motion for any system consonant with the principle of relativity if we equate the 4-vector $\frac{d\mathfrak{g}}{dt_0}$ to any other 4-vector.

This is the modified form of the facts contained in Newton's 1st and 2nd laws of motion.

The Newtonian 3rd law of course retains its meaning as far as *contact* action and reaction are concerned, since we maintain the principle of the conservation of momentum of aether and matter together, the magnitude of the action between them being none other than the rate of transfer of momentum (pp. 138 ff.). But this cannot be discussed further without entering into the obscure question of the nature of the action between aether and matter. The immediate purpose is to see what information can be gained as to the mechanical action of matter on matter, and here the *action* and *reaction* between two bodies are clearly not necessarily equal. So far the only knowledge we have is that the force on either body can be obtained from a 4-vector which, if the internal energy is unaltered, is orthogonal to its velocity 4-vector.

7. The inertia (Trägheit) of energy*.

We have had above an instance of a property which as will be seen below must, in the light of the principle of

* Cf. Larmor, "On the Dynamics of Radiation," *International Congress of Mathematicians*, Cambridge, 1912, pp. 213–6, especially the footnote, p. 216.

relativity, be taken as of general application, and of which another instance is well known and has been referred to in the preceding chapter.

The Poynting vector $\mathbf{Q}$ (p. 142) is equal to the momentum vector $[\mathbf{EH}]/c$ multiplied by c^2; that is, in the aether theory a 'momentum' $\mathbf{g}$ is accompanied by a 'flow of energy' $c^2\mathbf{g}$ in the same direction*.

In the particle dynamics which has been developed above we have the vector relation

$$c^2\mathbf{g} = w\mathbf{v}.$$

Now in this case the energy w is convected with the particle with velocity $\mathbf{v}$ so that there is a *flow of energy of amount equal to c^2 times the momentum* exactly as above (cf. § 6, p. 185).

8. The motion of a radiating system.

The above analysis is not restricted to a mere particle but applies equally well to the case of any system subject to the principle of relativity, the motion of which is purely translational.

There appears at first sight to be a contradiction between the general hypothesis of relativity and the results of ordinary theory in respect of such a system. If in a given frame of reference a self-contained system is radiating uniformly in all directions, it will clearly remain at rest if subject to no external influence. According to the principle of relativity therefore, if observed in another frame of reference, it will continue to move with uniform velocity.

Now it has been shewn on the older form of electromagnetic theory that a moving radiating body is subject to a resistance owing to its own radiation†. The suggestion is that owing to this resistance it will be retarded, contrary to the result anticipated by the hypothesis of relativity.

* From the point of view of the Principle of Relativity this is a sufficient reason for not considering other possible expressions for flow of energy.

† Cf. Larmor, *International Congress*, quoted above, p. 212, and Poynting, *Phil. Trans.* A, 202 (1903), p. 551.

But the reconciliation lies in the above proposition of the inertia of energy. According to that theorem, the uniformly moving radiating body is constantly losing energy and we have, since

$$\mathbf{g} = \frac{w\mathbf{v}}{c^2},$$

a corresponding loss of momentum given by

$$\frac{d\mathbf{g}}{dt} = \frac{\mathbf{v}}{c^2}\frac{dw}{dt}.$$

Thus what is in one theory spoken of as a resistance is in the other spoken of as loss of momentum, and there is no further contradiction.

A similar point occurs in connection with the Trouton-Noble experiment*, where ordinary theory predicts a couple acting on a suspended condenser when its energy is changed, whereas the Principle of Relativity demands that no rotation shall be observed—a demand satisfied in experiment. It can be shewn from the analysis to be given later, Chap. XIV, that the moment of momentum of a uniformly moving body is changed when its energy content is altered, and the rate of change is just equal to the couple predicted by ordinary theory. Thus the contradiction only arises in either case if we maintain the empirical dynamical assumptions of constant momentum and moment of momentum for a body moving uniformly with changing internal energy.

9. A kinetic potential.

The expressions obtained for $\mathbf{g}$ and w are related to a single function H as follows.

Putting $\quad H = -w_0/\kappa$

$$= -w_0\left(1 - \frac{v_x^2 + v_y^2 + v_z^2}{c^2}\right)^{\frac{1}{2}},$$

we have identically

$$g_x = \frac{\partial H}{\partial v_x},$$

* See Chap. IV, p. 39.

$$g_y = \frac{\partial H}{\partial v_y},$$

$$g_z = \frac{\partial H}{\partial v_z},$$

and
$$w = \Sigma\left(v\frac{\partial H}{\partial v}\right) - H,$$

and, if the internal energy is constant, the equations of motion become

$$\frac{d}{dt}\left(\frac{\partial H}{\partial v_r}\right) = k_r,$$

$$\frac{d}{dt}\left\{\Sigma\left(v\frac{\partial H}{\partial v}\right) - H\right\} = (\mathbf{kv}).$$

The analogy with classical dynamics is here very close.

Of this function H we may note the following property, that

$$H\delta t = -w_0\delta t_0;$$

that is, $H\delta t$ is an invariant, so that taking

$$A = \int H\,dt,$$

the equations of motion are given by the ordinary Hamiltonian method of variation in which we put

$$\delta A = \int \delta W\,dt,$$

where δU is the virtual work of the forces acting on the particle. The invariance of the form of the equations of motion is clearly shewn now by the fact that both sides of this variational equation are invariant.

10. Gravitation and the Principle of Relativity.

In seeking for phenomena which may throw some light on the mechanical theory that has been developed in connection with the principle of relativity, one is almost necessarily limited to the consideration of planetary motion. Terrestrial phenomena give relative velocities which are much too small to give any hope of any experimental comparison.

But the Newtonian law of gravitation has been so completely corroborated by the observed motions of the planets, that the statement has often been made, that if there is a finite velocity of propagation for gravitation it is vastly greater than that of light*.

Now the rise of an electromagnetic theory of matter shakes the foundations of this statement at two points. In the first place, if there is any justification for the hope that the electromagnetic theory may prove to be complete, gravitation would presumably come within its scope, though so far it has defied all attempts to include it. In this case we should anticipate its velocity of propagation to be that of light. In the second place the statement was based on the Newtonian conceptions of mass and force, and these have been seen to require some modification, whether in the manner above suggested or otherwise.

Without going into the entirely obscure question of the physical nature of gravitation, it is desirable therefore to see whether the equations for planetary motion, and incidentally the law of gravitation, can be so modified as to be consistent with the principle of relativity, while still remaining consistent with the observed facts of planetary motion.

It is clear that *the Newtonian law will not fit into the scheme of particle dynamics developed above.*

For, if the force upon a particle at rest due to the attraction of another is given by

$$\mathbf{k} = \gamma m_1 m_2 \left(\frac{x}{r^3}, \frac{y}{r^3}, \frac{z}{r^3}\right),$$

where $x = x_2 - x_1$, $y = y_2 - y_1$, $z = z_2 - z_1$, the respective positions being (x_1, y_1, z_1), (x_2, y_2, z_2) the extended force 4-vector for one of the particles would be

$$\mathfrak{k} = (\mathbf{k}, 0).$$

* For the possibility of small corrections to the Newtonian law and of the hypothesis of a finite velocity of propagation see Zenneck, *Enzyk. der Math. Wiss.* Vol. v. pp. 35 ff.

Applying a Lorentz transformation, so as to assign a velocity v to the particle along the axis of x, we should have

$$\mathfrak{k}' = \left(\beta k_x, k_y, k_z, -\frac{\beta v}{c^2} k_x\right)$$
$$= \gamma m_1 m_2 \left(\beta^2 \frac{x_2' - x_1'}{r^3}, \frac{y'}{r^3}, \frac{z'}{r^3}, -\frac{\beta^2 v (x_2' - x_1')}{c^2 r^3}\right),$$

where $r^2 = \beta^2 x'^2 + y'^2 + z'^2$, (x', y', z') being the differences of the simultaneous positions in the new coordinates, so that the new force in the Newtonian sense would be

$$\mathbf{k}' = \gamma m_1 m_2 \left(\frac{\beta x'}{r^3}, \frac{y'}{r^3}, \frac{z'}{r^3}\right),$$

which, even in this simple case where the particles have the same velocity, is quite different from that given by the Newtonian law, viz.

$$\mathbf{k}' = \gamma m_1 m_2 \left(\frac{x'}{r'^3}, \frac{y'}{r'^3}, \frac{z'}{r'^3}\right),$$

where $$r'^2 = x'^2 + y'^2 + z'^2.$$

11. Investigation of an invariant form of the equations of planetary motion.

The problem of finding an invariant form of the equations of motion was first attacked by Poincaré*, and incidentally he was the first to recognize the importance of the 4-vector as a means of representing the way in which a mechanical force must be transformed in the correlation of moving and stationary systems initiated by Lorentz, thus completing that discussion at an important point. What follows is essentially due to him.

Let A and B be two moving particles whose coordinates in a certain system of reference are (x_1, y_1, z_1), (x_2, y_2, z_2) at times t_1 and t_2 respectively. Calling their velocities $\mathbf{v}_1$ and $\mathbf{v}_2$ respectively, we proceed to examine the possibilities of specifying a

* *Rendiconti del Circ. mat. di Palermo*, 21 (1906), p. 166, "Sur la dynamique de l'électron."

force 4-vector depending on the coordinates and the velocities of the two particles and reducing for the case of particles at rest to the 4-vector

$$\mu\left(\frac{x}{r^3}, \frac{y}{r^3}, \frac{z}{r^3}, 0\right).$$

We must remember too that the force exerted by A on B is not now necessarily to be assumed equal and opposite to that exerted by B on A.

12. The law of propagation.

In order to eliminate a discussion of the mode of transmission of gravitation we must *assume a law of propagation*; that is, we assume that the force 4-vector for A at time t_1 depends on the position and motion of B at time t_2, t_2 being supposed to be in some fixed relation to t_1 and the motions of the two particles.

To maintain the principle of relativity we must assume that *this relation is an invariant one*, and, though this is not the only possible assumption, we shall consider what is the result of taking the relation

$$r = c(t_1 - t_2),$$

where r is the distance between the positions of the particles at times t_2, t_1 respectively; that is, we assume the influence of gravitation to be propagated with the velocity of light.

In dealing with the force on B due to A we should of course take

$$r = c(t_2 - t_1).$$

13. The introduction of 4-vectors and invariants.

We may now write down a number of 4-vectors which depend on the relative coordinates

$$(x_2 - x_1,\ y_2 - y_1,\ z_2 - z_1,\ t_2 - t_1)$$

and on $\mathbf{v}_1$ and $\mathbf{v}_2$.

The simplest are

$$\mathfrak{v}_1 = \kappa_1\left(\frac{\mathbf{v}_1}{c}, i\right) \ldots\ldots \text{ where } \kappa_1 = (1 - v_1^2/c^2)^{-\frac{1}{2}},$$

$$\mathfrak{v}_2 = \kappa_2\left(\frac{\mathbf{v}_2}{c}, i\right) \ldots\ldots \text{ where } \kappa_2 = (1 - v_2^2/c^2)^{-\frac{1}{2}},$$

$$\mathfrak{r} = \mathfrak{r}_1 - \mathfrak{r}_2 = \{x_1 - x_2, y_1 - y_2, z_1 - z_2, ic(t_1 - t_2)\}.$$

We shall not consider any quantities depending on the acceleration or higher derivatives of the velocities.

Then since a 4-vector multiplied by an invariant scalar is still a 4-vector, we can by taking invariant multiples of the above quantities obtain a very general 4-vector,

$$k = I\mathfrak{r} + I_1\mathfrak{v}_1 + I_2\mathfrak{v}_2,$$

where I, I_1, I_2 are any invariant functions of the coordinates and velocities*.

We have next to consider what invariant functions we may take for I, I_1, I_2. The three products

$$\rho_1 = (\mathfrak{v}_1\mathfrak{r}), \quad \rho_2 = (\mathfrak{v}_2\mathfrak{r}), \quad \gamma = (\mathfrak{v}_1\mathfrak{v}_2)$$

are the simplest and most fundamental, and any function of these will be an invariant. Thus we have the general type of 4-vector

$$\mathfrak{k} = f(\rho_1, \rho_2, \gamma)\,\mathfrak{r} + f_1(\rho_1, \rho_2, \gamma)\,\mathfrak{v}_1 + f_2(\rho_1, \rho_2, \gamma)\,\mathfrak{v}_2^*.$$

Now, with the restriction $r = c(t_1 - t_2)$, we have, putting $\mathbf{v}_1/c = (\xi_1, \eta_1, \zeta_1)$, and $(x, y, z) = (x_1 - x_2, y_1 - y_2, z_1 - z_2)$,

$$\rho_1 = -\kappa_1\{r - (\xi_1 x + \eta_1 y + \zeta_1 z)\},$$

and similarly $\rho_2 = -\kappa_2\{r - (\xi_2 x + \eta_2 y + \zeta_2 z)\}$,

and $\gamma = -\kappa_1\kappa_2\{1 - (\xi_1\xi_2 + \eta_1\eta_2 + \zeta_1\zeta_2)\}$;

so that, as $\mathbf{v}_1$ tends to zero, ρ_1 tends to r, and r tends to become the instantaneous distance between the particles; also, as $\mathbf{v}_2$ tends to zero, ρ_2 tends to r, while, when both $\mathbf{v}_1$ and $\mathbf{v}_2$ tend to zero, $-\gamma$ tends to unity.

* Another term might be added $I_3[\mathfrak{r}[\mathfrak{v}_1\mathfrak{v}_2]]$. If $\mathfrak{r}$, $\mathfrak{v}_1$, $\mathfrak{v}_2$ are in different directions, this vector is orthogonal to them all, and any 4-vector whatever could be expressed as a sum of the four vectors multiplied by invariant factors. This however is not considered by Poincaré.

Thus if we take

$$\mathfrak{k} = f\left(\frac{\rho_1}{\rho_2}, \gamma\right)\frac{\mathfrak{r}}{\rho_2^3} + f_1(\rho_1, \rho_2, \gamma)\,\mathfrak{v}_1 + f_2(\rho_1, \rho_2, \gamma)\,\mathfrak{v}_2,$$

we have a 4-vector whose first three components tend to the form

$$\mu(x, y, z)/r^3$$

as $\mathbf{v}_1$ and $\mathbf{v}_2$ tend to zero, since then

$$f\left(\frac{\rho_1}{\rho_2}, \gamma\right) = f(1, -1),$$

and is therefore a mere constant.

14. Invariant equation of gravitational motion.

Thus the equation

$$\frac{d\mathfrak{g}}{dt_0} = \mathfrak{k}$$

for the particle A becomes

$$\begin{aligned}\kappa_1 \frac{d}{dt}(\kappa_1 m_1 \mathbf{v}_1) = f\left(\frac{\rho_1}{\rho_2}, \gamma\right) & \frac{(x, y, z)}{\kappa_2^3\{r - (\xi_2 x + \eta_2 y + \zeta_2 z)\}^3} \\ & + f_1(\rho_1, \rho_2, \gamma)\,\mathbf{v}_1 \\ & + f_2(\rho_1, \rho_2, \gamma)\,\mathbf{v}_2 \quad \ldots\ldots\ldots\ldots\ldots\ldots\ldots(\alpha).\end{aligned}$$

This general form is consistent with the principle of relativity, and it remains to consider whether any particular case of it is consistent with the facts of observation. If it is so, then the objection to the finite velocity of propagation of gravitation is removed, and the principle of relativity is shewn to be not inconsistent with gravitational phenomena.

In restricting the form of (α) we may note first that if the internal state of the attracted body is unchanged by the attraction, $\mathfrak{k}$ is orthogonal to $\mathfrak{v}_1$ (p. 167).

Thus

$$\begin{aligned} 0 = (\mathfrak{k}\mathfrak{v}_1) &= f\left(\frac{\rho_1}{\rho_2}, \gamma\right)\frac{(\mathfrak{v}_1\mathfrak{r})}{\rho_2^3} + f_1(\rho_1, \rho_2, \gamma)\,\mathfrak{v}_1^2 + f_2(\rho_1, \rho_2, \gamma)\,(\mathfrak{v}_1\mathfrak{v}_2) \\ &= f\left(\frac{\rho_1}{\rho_2}, \gamma\right)\frac{\rho_1}{\rho_2^3} - f_1(\rho_1, \rho_2, \gamma) + f_2(\rho_1, \rho_2, \gamma)\,\gamma. \end{aligned}$$

This is to be identically true.

Poincaré satisfies it by putting

$$f_1(\rho_1, \rho_2, \gamma) \equiv 0,$$

and therefore

$$f_2(\rho_1, \rho_2, \gamma) \equiv -\frac{1}{\rho_1^2\gamma} f\left(\frac{\rho_1}{\rho_2}, \gamma\right),$$

so that
$$\mathfrak{k} = f\left(\frac{\rho_1}{\rho_2}, \gamma\right)\left\{\frac{\mathfrak{r}}{\rho_2^3} - \frac{\rho_1\mathfrak{b}_2}{\gamma\rho_2^3}\right\},$$

of which the simplest particular case is

$$\mathfrak{k} = \frac{\mu}{\rho_2^3}\left\{\mathfrak{r} - \frac{\rho_1\mathfrak{b}_2}{\gamma}\right\},$$

where μ is a constant; giving a Newtonian force

$$\mathbf{k} = \frac{\mu}{\kappa_1\rho_2^3}\left\{(x, y, z) - \frac{\rho_1\mathbf{v}_2}{\gamma c}\right\}.$$

Now, again to the first order, if (x', y', z') are the coordinates of the position of A at time t_1 relative to the *simultaneous* position of B, we have

$$(x', y', z') = (x, y, z) - (t_1 - t_2)\,\mathbf{v}_2$$
$$= (x, y, z) - \frac{r}{c}\mathbf{v}_2.$$

If
$$r_1'^2 = x'^2 + y'^2 + z'^2,$$

so that r_1' is the distance of the simultaneous positions of A and B at time t_1, then to the first order of small quantities

$$r_1'^2 = r^2 - 2r(x\xi_2 + y\eta_2 + z\zeta_2),$$

or
$$r_1' = r\left\{1 - \frac{x\xi_2 + y\eta_2 + z\zeta_2}{r}\right\}$$
$$= \rho_2,$$

so that to this order

$$\mathbf{k} = \mu\left\{-\frac{(x', y', z')}{r_1'^3} - \frac{\mathbf{v}_2}{cr_1'^2} + \frac{\mathbf{v}_2}{cr_1'^2}\right\},$$

since, neglecting $\mathbf{v}_1$ and $\mathbf{v}_2$,

$$\frac{\rho_1}{\gamma\rho_2^3} = -\frac{1}{r_1'^2}.$$

Thus to this order

$$\mathbf{k} = -\mu\left(\frac{x', y', z'}{r_1'^3}\right).$$

15. Reduction of equations to Newtonian form to the first order.

For simplicity, let us consider the case in which the velocity of B does not change, so that we may take a system of reference in which it is at rest. Then to the first order of small quantities in this system the equation (α), p. 176, becomes

$$m_1 \frac{d^2}{dt^2}(x', y', z') = -\mu\left(\frac{x', y', z'}{r'^3}\right),$$

which are the ordinary equations of motion of Newtonian dynamics.

In the same way, in the more general case when A and B are both accelerated, the equations

$$m_1 \frac{d^2(x_1, y_1, z_1)}{dt^2} = -\mu\left(\frac{x', y', z'}{r'^3}\right),$$

$$m_2 \frac{d^2(x_2, y_2, z_2)'}{dt^2} = +\mu\left(\frac{x', y', z'}{r'}\right),$$

(x_1, y_1, z_1), $(x_2, y_2, z_2)'$ being *simultaneous* coordinates, are consistent with the principle of relativity to the first order of small quantities. Here, we remember, (x', y', z') are simultaneous relative coordinates.

16. The second order corrections inappreciable.

The possibility of obtaining equations which, to the first order, are of Newtonian form removes the old objection to the velocity of propagation of gravitation being c, an objection which was based on the prediction of a *first order* effect.

But for a complete comparison with astronomical observations it is necessary to examine the nature and magnitude of the second order effect. This has been carefully and exhaustively done by Professor de Sitter*. It would carry us too far to give the calculations here, but the results may be summarized.

* *Monthly Notices of Roy. Astr. Soc.* Mar. 1911, p. 388.

Taking the following equations, either of which is a particular case of (α), p. 176,

$$m_1 \frac{d^2(x_1, y_1, z_1)}{dt_0^2} = \frac{\mu}{\rho_2^3}\left\{(x, y, z) - \frac{\rho_1 \mathbf{v}_2}{\gamma c}\right\} \quad \text{...........(I)},$$

and

$$m_2 \frac{d^2(x_1, y_1, z_1)}{dt_0^2} = \frac{\mu(-\gamma)}{\rho_2^3}\left\{(x, y, z) - \frac{\rho_1 \mathbf{v}_2}{\gamma c}\right\} \quad \text{......(II)},$$

(II) differing from (I) only in the extra invariant factor $(-\gamma)$ on the right-hand side—de Sitter approximates to the second order in both cases and comes to the following conclusions.

Case I.

(i) The coordinates of a planet of small mass are expressed by the ordinary formulae of elliptical motion.

(ii) But to express the eccentric anomaly in terms of the heliocentric time we must take a slightly altered eccentricity, the difference between heliocentric and geocentric time consisting in a small change of scale together with small periodic fluctuations.

(iii) Kepler's third law is not quite exact, but there are periodic variations.

(iv) The difference between the constant of precession as determined from the fixed stars and from the motions in the solar system would be of the order of

$$-0''\cdot0000044 \text{ per century.}$$

The variation in the eccentricity in (ii) is of the order $(v/c)^2$ of itself, and for the earth this is of the order 10^{-8}.

The periodic change in the time in (ii) has amplitude $(v^2 e/c^2 n)$, n being the mean angular velocity, and is approximately equal to 0·0008 second.

The deviation from the Keplerian angular velocity in (iii) is again of the order $(v/c)^2$ of the mean, that is of the order 10^{-8}.

All these effects are inappreciable.

There is really no need to go any further, as these results, if correct, shew that *there is no essential inconsistency between astronomical observations and the Principle of Relativity.*

De Sitter however goes on to shew that the equation (II) also leads to results which are at present incapable of observation, except in one important respect. He finds in fact that this equation would lead to a secular motion of the perihelia of the planets which in the case of Mercury amounts to about 7″ per century. An effect of this kind has for some time been known by practical astronomers to exist, though the magnitude is about 40″ per century. Various hypotheses have been suggested to explain it. One of them proposed by Gerber* in 1898 quite independently of the principle of relativity is the possibility that the Newtonian Law of Gravitation is only approximate, and that more accurately gravitational influence is propagated with the velocity of light, and that a correction of nature very similar to that suggested by equation (II) must be applied to the usual expression for the force on the planet. He arrives at the conclusion that the known motion of the perihelia can be so explained.

By using instead of equation (II) an equation derived from (I) by multiplying the right-hand side by another power of the invariant factor $(-\gamma)$ instead of the first, the magnitude of the effect predicted could be made just of the actual order, and Gerber's conclusion is thereby corroborated and found to be perfectly consistent with the hypothesis of relativity.

* *Zeitschr. für Math. Phys.* 43 (1908), pp. 93–104. See also *Enzyk. der Math. Wiss.* Vol. v. p. 49.

CHAPTER XIV

THE DYNAMICS OF CONTINUOUS MATERIAL MEDIA*

1. The ordinary equations of motion of a material media.

At constant temperature these are of the form†

$$\frac{\partial p_{xx}}{\partial x} + \frac{\partial p_{xy}}{\partial y} + \frac{\partial p_{xz}}{\partial z} + \frac{D}{\partial t}(\rho v_x) = k_x,$$

$$\frac{\partial p_{yx}}{\partial x} + \frac{\partial p_{yy}}{\partial y} + \frac{\partial p_{yz}}{\partial z} + \frac{D}{\partial t}(\rho v_y) = k_y,$$

$$\frac{\partial p_{zx}}{\partial x} + \frac{\partial p_{zy}}{\partial y} + \frac{\partial p_{zz}}{\partial z} + \frac{D}{\partial t}(\rho v_z) = k_z,$$

and the energy equation derived from them is

$$\frac{\partial}{\partial x}(v_x p_{xx} + v_y p_{xy} + v_z p_{xz}) + \frac{\partial}{\partial y}(v_x p_{yx} + v_y p_{yy} + v_z p_{yz})$$
$$+ \frac{\partial}{\partial z}(v_x p_{zx} + v_y p_{zy} + v_z p_{zz})$$
$$+ \frac{D}{\partial t}\{\tfrac{1}{2}\rho(v_x^2 + v_y^2 + v_z^2)\} + V = k_x v_x + k_y v_y + k_z v_z,$$

where V is the potential energy per unit volume, and k_x, k_y, k_z the body force per unit volume.

Here (k_x, k_y, k_z), (v_x, v_y, v_z) are Newtonian vectors, $p_{xy} = p_{yx}$

* This chapter might be omitted by non-mathematical readers.

† Where $\dfrac{D\phi}{\partial t} \equiv \dfrac{\partial \phi}{\partial t} + \dfrac{\partial}{\partial x}(\phi v_x) + \dfrac{\partial}{\partial y}(\phi v_y) + \dfrac{\partial}{\partial z}(\phi v_z)$.

and $(p_{xx}, p_{yy}, p_{zz}, p_{yz}, p_{zx}, p_{xy})$ form a 'tensor'*; the characteristic property of a tensor is that, (ξ, η, ζ) being any vector,

$$\xi p_{xx} + \eta p_{xy} + \zeta p_{xz},$$
$$\xi p_{yx} + \eta p_{yy} + \zeta p_{yz},$$
$$\xi p_{zx} + \eta p_{zy} + \zeta p_{zz}$$

form another vector.

Now the above equations may be written

$$\frac{\partial q_{xx}}{\partial x} + \frac{\partial q_{xy}}{\partial y} + \frac{\partial q_{xz}}{\partial z} + \frac{\partial g_x}{\partial t} = k_x,$$

$$\frac{\partial q_{yx}}{\partial x} + \frac{\partial q_{yy}}{\partial y} + \frac{\partial q_{yz}}{\partial z} + \frac{\partial g_y}{\partial t} = k_y,$$

$$\frac{\partial q_{zx}}{\partial x} + \frac{\partial q_{zy}}{\partial y} + \frac{\partial q_{zz}}{\partial z} + \frac{\partial g_z}{\partial t} = k_z,$$

together with

$$\frac{\partial Q_x}{\partial x} + \frac{\partial Q_y}{\partial y} + \frac{\partial Q_z}{\partial z} + \frac{\partial w}{\partial t} = (k_x v_x + k_y v_y + k_z v_z),$$

where $\mathbf{g} = (g_x, g_y, g_z) = \rho\,(v_x, v_y, v_z)$ is the momentum per unit volume, and

$$q_{xx} = p_{xx} + v_x g_x,$$
$$q_{xy} = p_{xy} + v_y g_x,$$
$$q_{yx} = p_{yx} + v_x g_y, \text{ etc.}$$

and $\mathbf{Q} = (Q_x, Q_y, Q_z)$ denotes the flux of energy relative to a fixed frame of reference.

2. The condition of relativity.

For the sake of symmetry we will write

$$icg_x = q_{xu}, \quad icg_y = q_{yu}, \quad icg_z = q_{zu},$$

$$\frac{i}{c} Q_x = q_{ux}, \quad \frac{i}{c} Q_y = q_{uy}, \quad \frac{i}{c} Q_z = q_{uz},$$

$$k_u = \frac{i}{c}(k_x v_x + k_y v_y + k_z v_z),$$

and $$-w = q_{uu}.$$

* See p. 140, footnote.

Then the four equations become

$$\frac{\partial q_{xx}}{\partial x} + \frac{\partial q_{xy}}{\partial y} + \frac{\partial q_{xz}}{\partial z} + \frac{\partial q_{xu}}{\partial u} = k_x,$$

$$\frac{\partial q_{yx}}{\partial x} + \frac{\partial q_{yy}}{\partial y} + \frac{\partial q_{yz}}{\partial z} + \frac{\partial q_{yu}}{\partial u} = k_y,$$

$$\frac{\partial q_{zx}}{\partial x} + \frac{\partial q_{zy}}{\partial y} + \frac{\partial q_{zz}}{\partial z} + \frac{\partial q_{zu}}{\partial u} = k_z,$$

$$\frac{\partial q_{ux}}{\partial x} + \frac{\partial q_{uy}}{\partial y} + \frac{\partial q_{uz}}{\partial z} + \frac{\partial q_{uu}}{\partial u} = k_u.$$

We will denote this equation by

$$\operatorname{div} \mathfrak{Q} = \mathfrak{k}.$$

These equations will be of invariant form provided that $\mathfrak{k} = (k_x, k_y, k_z, k_u)$ *is a 4-vector, and also provided that the left-hand sides of the four equations are the components of a 4-vector.*

The left-hand side of the above equations is in form an exact extension to four dimensions of the equations of equilibrium of an elastic medium in three dimensions. We shall accordingly call $\mathfrak{Q}$ a 'generalized tensor' (see p. 140, footnote).

3. The requirement that $\mathfrak{k}$ shall be a 4-vector is exactly that which arises in the electromagnetic theory for the force per unit volume on an electric charge (see p. 159). These two facts are necessarily related.

In fact in a system whose motion is determined by the assumption that the electromagnetic forces and mechanical forces balance at every point we must have

$$(k_x, k_y, k_z)_1 = -(k_x, k_y, k_z)_2,$$

where the suffix 1 refers to the electromagnetic and the suffix 2 to the mechanical force. Introducing the rate of work of the force as above, we have therefore

$$(k_x, k_y, k_z, k_u)_1 = -(k_x, k_y, k_z, k_u)_2.$$

If therefore the left-hand side is a 4-vector, the right-hand side must be a 4-vector also.

Thus *the introduction of the hypothesis of equal and opposite action and reaction as regards mechanical and electromagnetic forces requires the mechanical forces to obey the transformation required by the principle of relativity.*

4. The transformation of the generalized tensor $\mathfrak{Q}$.

Turning to the left-hand side of the equation, we know that the operator $\left(\frac{\partial}{\partial x}, \frac{\partial}{\partial y}, \frac{\partial}{\partial z}, \frac{\partial}{\partial u}\right)$ is subject to the same transformation as any 4-vector.

Hence the condition that div $\mathfrak{Q}$ shall be a 4-vector is that, *if* $\mathfrak{a}$ *is any* 4-*vector* $(\mathfrak{a}\mathfrak{Q})$ *shall be a* 4-*vector*, where $(\mathfrak{a}\mathfrak{Q})_x$ stands for

$$a_x q_{xx} + a_y q_{xy} + a_z q_{xz} + a_u q_{xu},$$

and so for the other components.

Hence $(\mathfrak{b}\,(\mathfrak{a}\mathfrak{Q}))$ must be an invariant, where $\mathfrak{b}$ is any other 4-vector, that is

$$q_{xx} a_x b_x + q_{xy} b_x a_y + q_{yx} b_y a_x + \dots$$

must be invariant.

Hence the formula of transformation for the generalized tensor $\mathfrak{Q}$ can be written down, by taking special values for the 4-vectors $\mathfrak{a}$, $\mathfrak{b}$, such as (1, 0, 0, 0), (0, 1, 0, 0), etc.

Further, since

$$(\mathfrak{a}\mathfrak{c})(\mathfrak{b}\mathfrak{d}) = c_x d_x \,.\, a_x b_x + c_y d_x \,.\, a_y b_x + c_x d_y \,.\, a_x b_y + \dots$$

is invariant, whatever the 4-vectors $\mathfrak{c}$, $\mathfrak{d}$, the transformation for $\mathfrak{Q}$ must be the same as that for the array

$$\begin{pmatrix} c_x d_x, & c_y d_x, & c_z d_x, & c_u d_x \\ c_x d_y, & c_y d_y, & c_z d_y, & c_u d_y \\ c_x d_z, & c_y d_z, & c_z d_z, & c_u d_z \\ c_x d_u, & c_y d_u, & c_z d_u, & c_u d_u \end{pmatrix}.$$

5. Application of the transformation of $\mathfrak{Q}$.

Suppose now that we make a further assumption that to *an observer at rest with any point of the system, the momentum-density and also the flow of energy at that point is zero.* This would preclude such transfer of energy as is involved in a flow of heat; we may call the motion *adiabatic*.

Then starting from the equation div $\mathfrak{Q} = \mathfrak{k}$, if we change the frame of reference to one in which any particular point of the body is at rest, we have

$$q_{xu}' = q_{yu}' = q_{zu}' = q_{ux}' = q_{uy}' = q_{uz}' = 0$$

at that point.

Further, in the ordinary material mechanics we have

$$p_{yz} = p_{zy}, \quad p_{zx} = p_{xz}, \quad p_{xy} = p_{yx},$$

which in the case of a stationary point of a body are the same as

$$q_{yz} = q_{zy}, \quad q_{zx} = q_{xz}, \quad q_{xy} = q_{yx}.$$

If we transfer this assumption for a body at rest to our case, the tensor $\mathfrak{Q}'$ becomes a symmetrical one for that frame of reference in which the point in question is at rest.

If therefore by any transformation, $\mathfrak{a}$, $\mathfrak{b}$, $\mathfrak{Q}$ are transformed into $\mathfrak{a}'$, $\mathfrak{b}'$, $\mathfrak{Q}'$, since the invariant $(\mathfrak{b}'(\mathfrak{a}'\mathfrak{Q}'))$ is symmetrical in $\mathfrak{a}'$ and $\mathfrak{b}'$, so is the invariant $(\mathfrak{b}(\mathfrak{a}\mathfrak{Q}))$ symmetrical in $\mathfrak{a}$ and $\mathfrak{b}$.

Thus $$q_{xu} = q_{ux},$$

or $$\frac{i}{c} Q_x = icg_x,$$

that is $$Q_x = c^2 g_x,$$

and similarly $$Q_y = c^2 g_y,$$

and $$Q_z = c^2 g_z,$$

or $$\mathbf{Q} = c^2 \mathbf{g}.$$

6. The inertia of energy.

We now see what is the actual significance of the theorem referred to above (p. 169) that a flux of energy $\mathbf{Q}$ implies a momentum $\mathbf{Q}/c^2$.

The assumptions from which this has been deduced in the preceding section are

(i) The hypothesis of relativity;

(ii) The vanishing of the momentum and of transfer of energy at a point which is at rest;

(iii) The symmetry of the ordinary stress tensor at a point which is at rest.

The third assumption we know follows in ordinary theory from the statical theorem of moments.

7. The stress in the medium.

In what has been said above the quantities

$$(q_{xx}, q_{yy}, q_{zz}, q_{yz}, q_{zx}, q_{xy})$$

are not what are properly called *stresses*, inasmuch as they only give the rate of transfer of momentum *relative to the system of axes*, and not relative to the medium itself.

The true stresses are the quantities

$$p_{xx} = q_{xx} - v_x g_x,$$

$$p_{xy} = q_{xy} - v_y g_x, \text{ etc.,}$$

which are the rates of transfer of momentum *relative to the body itself*. Now in ordinary material mechanics, **g** is a vector in the same direction as **v**, so that

$$v_x g_y = v_y g_x, \text{ etc.}$$

In that case therefore the 3-tensor p_{xx}, etc., is also symmetrical.

8. The momentum not in the direction of the velocity in a strained medium, except in the case of uniform pressure.

But the momentum **g** in our case is not an ordinary vector, in fact the transformation obtained above for the tensor $\mathfrak{Q}$ shews that **g** is not in general in the direction of the velocity **v**.

To see this, let us apply the Lorentz transformation to the tensor $\mathfrak{Q}$ for a point at rest to obtain the tensor $\mathfrak{Q}'$ for the same point moving with uniform velocity v parallel to the axis of x, the transformation for a 4-vector being written

$$x' = Ax + Bu, \quad y' = y, \quad z' = z, \quad u' = Au - Bx,$$

where $$A = \frac{1}{(1 - v^2/c^2)^{\frac{1}{2}}}, \quad B = \frac{-iv/c}{(1 - v^2/c^2)^{\frac{1}{2}}}.$$

Then $\mathfrak{c}$, $\mathfrak{d}$ being any 4-vectors transforming into $\mathfrak{c}'$, $\mathfrak{d}'$,

$$c_u' d_x' = (A c_u - B c_x)(A d_x + B d_u).$$

Hence from § 4

$$q_{xu}' = -AB(q_{xx} - q_{uu}) - B^2 q_{ux} + A^2 q_{xu}$$
$$= -AB(q_{xx} - q_{uu}),$$

since $$q_{ux} = q_{xu} = 0.$$

Thus, remembering that

$$g_x' = \frac{q_{xu}'}{ic},$$

we have $$g_x' = \frac{\beta^2 v}{c^2}(q_{xx} - q_{uu})$$
$$= \frac{\beta^2 v}{c^2}(w + q_{xx}).$$

Again $$q_{yu}' = A q_{yu} - B q_{yx}$$
$$= -B q_{yx},$$

giving $$g_y' = \frac{\beta v}{c^2} q_{yx},$$

and similarly $$g_z' = \frac{\beta v}{c^2} q_{zx}.$$

Thus the momentum $\mathbf{g}$ is only in the direction of the velocity v, that is, in the direction of the axis of x, if the two stress components q_{yx} and q_{zx} are zero: and if this is to be so whatever the direction of the axes of (x, y, z), the state of stress for the body at rest becomes a simple hydrostatic pressure.

9. The transformation of the true stresses.

Remembering that

$$p_{xx} = q_{xx} - v_x g_x = q_{xx} + \frac{iv}{c} q_{xu},$$

and so on, we may write down the following scheme of transformation for the true stresses:

$$\begin{aligned}
p_{xx}' &= q_{xx}' + \frac{iv}{c} q_{xu}' \\
&= A^2 q_{xx} + B^2 q_{uu} + B^2 (q_{xx} - q_{uu}) \\
&= (A^2 + B^2) q_{xx} \\
&= q_{xx}, \\
p_{xy}' &= q_{xy}' + \frac{iv}{c} q_{yu}' \\
&= A q_{xy} + \frac{B^2}{A} q_{xy} \\
&= \frac{q_{xy}}{A}, \\
p_{yx}' &= q_{yx}' \\
&= A q_{yx}, \\
p_{yy}' &= q_{yy},
\end{aligned}$$

and so on.

This stress system is only symmetrical in the case referred to at the end of the last section, where the stress q_{xx}, etc., reduces to a uniform pressure, so that **g** and **v** are in the same direction.

10. The case of hydrostatic pressure; p an invariant.

In the particular case when the stress reduces to a uniform pressure in all directions, that is, when for the system at rest

$$q_{xx} = q_{yy} = q_{zz}, \quad q_{xy} = q_{yz} = q_{zx} = 0,$$

we have from the equations of the last section

$$p_{xx} = p_{yy} = p_{zz} = q_{xx}, \quad p_{yz} = p_{zx} = p_{xy} = 0.$$

Thus for the moving system also the stress reduces to the same simple type, and the magnitude of the pressure is an invariant.

Calling this p, and w_0 the energy per unit volume in the system at rest, since

$$q_{uu} = - w_0,$$

we have

$$w = q_{uu}' = - A^2 q_{uu} - B^2 q_{xx} = \beta^2 \left(w_0 + \frac{v^2}{c^2} p \right),$$

and

$$\mathbf{g} = \frac{\beta^2 \mathbf{v}}{c^2} (w_0 + p).$$

11. Illustration of the application of these formulae; the Lorentz electron.

If a charge q is distributed uniformly over a sphere of radius a, this produces an outward tension on the substance of the sphere. A sufficient means of balancing this outward tension would be to imagine the sphere to be fluid with a constant negative hydrostatic pressure at all interior points of magnitude equal to the force per unit area of the surface of the sphere due to the charge.

In the units we have been using, this is

$$p = - \tfrac{1}{2} \sigma^2 = - \frac{e^2}{32 \pi^2 a^4}.$$

The total energy of the sphere when at rest is

$$E_0 = \tfrac{1}{2} e V = \frac{e^2}{8 \pi a}.$$

Thus in the transformed system we shall have for points in the interior

$$\mathbf{g} = - \beta^2 \frac{\mathbf{v}}{c^2} \cdot \frac{e^2}{32 \pi^2 a^4},$$

giving a total momentum for the whole volume, $(\frac{4}{3} \pi a^3 / \beta)$, of amount

$$- \beta \frac{\mathbf{v}}{c^2} \cdot \frac{e^2}{24 \pi a^3}.$$

Consider next the shell of infinitesimal thickness which carries the charge. In traversing this the pressure p will rapidly diminish to zero, so that the contribution $\beta^2 \mathbf{v} p / c^2$ to the density of momentum will remain finite and therefore, for the infinitesimal volume of the layer, give a total momentum of vanishing amount.

On the other hand w_0 the density of energy is very large, the total energy being finite.

Thus the total momentum arising from the shell will be

$$(V_0/\beta)(\beta^2 w_0 v/c^2),$$

where V_0 is the volume of the shell when at rest, and

$$V_0 w_0 = W_0 = \frac{e^2}{8\pi a}.$$

Thus the momentum from this source is

$$\mathbf{g}_1 = -\tfrac{1}{3} W_0 \frac{\beta \mathbf{v}}{c^2}.$$

In the same way the energy $\beta^2 v^2 p/c^2$ per unit volume (§ 10) arising from this pressure gives a total for the whole volume V_0/β

$$\tfrac{4}{3}\pi a^3 \frac{\beta v^2}{c^2} p = -\frac{\beta v^2}{c^2} \cdot \frac{e^2}{24\pi a}$$

$$= -\beta W_0 \frac{v^2}{3c^2}.$$

A calculation of the electromagnetic momentum and energy of the field (see p. 146) gives

$$\mathbf{g}_2 = \tfrac{4}{3} \frac{\beta \mathbf{v}}{c^2} W_0,$$

$$W_2 = \beta W_0 \left(1 + \frac{3v^2}{c^2}\right),$$

giving for the total

$$\mathbf{g}_1 + \mathbf{g}_2 = \beta W_0 \mathbf{v}/c^2,$$

$$W_1 + W_2 = \beta W_0,$$

results which are, as we should expect, identical with those obtained from a different standpoint in Chapter XI, pp. 145–7.

12. Dynamics as a generalized statics.

The analysis that has been given of the extension of the principle of relativity to cover the dynamics of a material medium shews even more clearly than the discussions in the preceding chapters how the interdependence of the measures of space and time is only one aspect of a far-reaching interdependence between the geometry of rest and motion, between static and kinetic phenomena. The form which has been given to the dynamical equations of a medium is an exact analogy of the ordinary statical equations.

With Minkowski space and time become particular aspects of a single four-dimensional concept; the distinction between them as separate modes of correlating and ordering phenomena is lost, and the motion of a point in time is represented as a stationary curve in four-dimensional space. Now if all motional phenomena are looked at from this point of view, they become timeless phenomena in four-dimensional space. The whole history of a physical system is laid out as a changeless whole.

The origin of this view of time and space, and of the whole theory of relativity, was, as has been shewn above, the possibility of putting the field equations of electrodynamics into an invariant vectorial form, and we have seen that this form was an exact generalization to the four-dimensional space of the fundamental equations of electrostatic phenomena in ordinary space, viz.

$$\operatorname{div} \mathbf{E} = \rho, \quad \operatorname{curl} \mathbf{E} = 0.$$

We have now seen how, in examining how far it is possible by a suitable modification to reconcile the usual laws of dynamical phenomena with this view of time and space, we are led to equations which are again a generalization of the statical equations of three dimensions, and we have seen that, in order to effect the generalization, we have been led to contemplate an interdependence between the ideas of stress, energy and momentum which does not exist in the Newtonian dynamics.

A similar generalization of a statical theorem to four dimensions was noted in Chapter IX in considering the relation of the electrodynamic equations to the principle of least action, and it is worth noting that the equations of this chapter can be obtained immediately from that principle, by assuming for the action a space-time integral of the generalization to four dimensions of the strain-energy function of ordinary elastic problems. This generalized function must be restricted to be invariant when the frame of reference is changed, and to reduce for the particular case of a body whose velocity is everywhere small to the usual expression of the ordinary theory. This has been considered by Herglotz*, but it would carry us too far to give details here. The same conclusion as to the *inertia of energy* arises again in this method of approaching the question.

* *Ann. der Phys.* 36 (1911).

CHAPTER XV

RELATIVITY AND AN OBJECTIVE AETHER

1. It is an outstanding objection to the principle of relativity that the propagation of electromagnetic effects through space seems inseparable in thought from the conception of an objective aether; whereas, if the principle is in fact universally valid, such a conception seems impossible, for the mind shrinks from an objective medium which is not in some sense unique.

It is the object of this chapter to examine this difficulty, not with the hope of finally settling the question, but of shewing that a reconciliation is not impossible.

In the first place, the objective aether as commonly now conceived is a medium everywhere at rest*—or approximately so—through which a disturbance is propagated according to certain laws, the disturbances constituting in effect the phenomena of which we are cognizant. But although the attempt to assimilate the medium to a stationary elastic solid of purely Newtonian mechanical properties has been given up, yet the influence of that phase in the development of the conception remains in that the aether is in practice identified with the frame of reference relative to which the motion of bodies is recorded.

* But compare the Address given by Sir Joseph Larmor to the International Congress of Mathematicians, Cambridge, 1912, on the Dynamics of Radiation, in which the possibilities of 'convected aethereal fields' are considered, some of the results there given being particular cases of the general discussion which follows. The conception of moving Faraday tubes developed by Sir J. J. Thomson, *Recent Researches in Electricity and Magnetism*, 1893, Chap. I, should also be compared with the suggestions as to the possibility of a conceptual moving aether which are made below. In particular cases the velocity is the same in the two conceptions.

In fact the conception is practically that of an ideal infinite rigid body with all its material properties removed and certain others substituted. But as was pointed out in Chapter III, pp. 28–30, the aether as an objective medium to which motion or rest must be ascribed has not yet been defined. The electromagnetic vectors have not yet been correlated with any state of motion of the aether.

2. A mechanical moving aether possible in the light of the principle of relativity.

But, as we have seen, in the recent development of electromagnetic theory certain mechanical terms have come back into use. We speak now of 'electromagnetic momentum,' 'energy,' 'flow of energy,' and 'stress' in the aether. But in general these ideas are not associated with a 'velocity' in the aether; no relation is set up between stress, velocity, and flow of energy, such as always exists in the mechanical transmission of energy.

The question arises, starting from any frame of reference for which the fundamental equations are satisfied, *Is it possible, by assigning a suitable velocity to the aether at all points, to make a state of stress in the medium account both for the transference of momentum, and for the flow of energy?* If this were possible we should be giving objectivity to the aether in the sense that we are defining its motion*.

It does in fact appear possible to do this†. The velocity which is necessary has a certain definite component in the direction of the momentum vector, but there is in addition to this an arbitrary component in a certain determined direction in the plane containing the electric and magnetic vectors.

But the point now arises, *Is such a moving aether any more reconcilable with the principle of relativity than the fixed and rigid aether?*

* Cf. the kinematical aether suggested by Larmor in which magnetic intensity is proportional to velocity, and electric intensity to rotational strain. *Aether and Matter*, Appendix E, pp. 322–337.

† v. *Proc. Roy. Soc.* Vol. 83, p. 110. The analysis which follows is an extension of that in this paper.

This question must be answered thus: *It is valid to think of a moving medium as having objective reality provided that its velocity conforms to the kinematics of the principle of relativity.* In other words, *the velocities attributed to the aether in two frames of reference must be related to one another by the Einstein addition equation for the transformation from one frame to the other.*

Can this condition be satisfied by the velocity of the aether as determined by the above mechanical conditions?

The answer to this question will be shewn to be in the affirmative *provided that the undetermined component is taken of such magnitude as to make the total velocity of the aether at every point the velocity of light.*

Further, when the analysis is carried out, the stress in the moving aether (i.e. the rate of transference of momentum per unit area across elements of area moving with the velocity of the medium) is found to reduce to a particularly simple form (a slight modification only of the Maxwell electrostatic stress in the sense necessary for a moving medium) *whose principal components are invariant under the transformations of the principle.*

3. Proof of the above statements.

We have seen in the last chapter that if we attribute a velocity $\mathbf{v}$ to the medium, the true stress is not given by the tensor as found for the rate of flow of momentum across fixed elements of area (p. 140), but by the tensor giving the flow across elements moving with the medium, that is the tensor, using the notation of Chapter XI,

$$-\begin{vmatrix} X_x + v_x g_x, & X_y + v_y g_x, & X_z + v_z g_x \\ Y_x + v_x g_y, & Y_y + v_y g_y, & Y_z + v_z g_y \\ Z_x + v_x g_z, & Z_y + v_y g_z, & Z_z + v_z g_z \end{vmatrix}.$$

The rate at which this stress does work on an element of area normal to any direction must, if we are to satisfy the principle of energy in the mechanical sense, be equal to the flow of energy across the element in that direction.

Taking an element to which the normal has direction cosines (l, m, n), this rate of work per unit area is

$$-(lU_x + mU_y + nU_z),$$

where

$$U_x = (X_x + v_x g_x)\, v_x + (Y_x + v_x g_y)\, v_y + (Z_x + v_x g_z)\, v_z.$$

Substituting the expressions for the stresses, this becomes

$$\begin{aligned} U_x = &- \{\tfrac{1}{2} v_x (E_x^2 + H_x^2 - E_y^2 - H_y^2 - E_z^2 - H_z^2) \\ &+ v_y (E_x E_y + H_x H_y) + v_z (E_x E_z + H_x H_z) + v_x (\mathbf{vg})\} \\ = &\; v_x \{w - (\mathbf{vg})\} - E_x (\mathbf{vE}) - H_x (\mathbf{vH}), \end{aligned}$$

or, generally, the flow is given by the vector

$$-\mathbf{v}\,\{w - (\mathbf{vg})\} + \mathbf{E}\,(\mathbf{vE}) + \mathbf{H}\,(\mathbf{vH}).$$

Now according to the analysis of the energy (p. 142) the flow of energy across this same element must be equal to

$$\mathbf{Q} - w\mathbf{v}.$$

Thus the following equation is to be satisfied if the stress is to account for the flow of energy

$$\mathbf{Q} - w\mathbf{v} = \mathbf{v}\,\{w - (\mathbf{vg})\} - \mathbf{E}\,(\mathbf{vE}) - \mathbf{H}\,(\mathbf{vH}),$$

or

$$c^2\mathbf{g} + \mathbf{E}\,(\mathbf{vE}) + \mathbf{H}\,(\mathbf{vH}) = \mathbf{v}\,\{2w - (\mathbf{vg})\} \quad \text{......(1)}.$$

We have seen in the last chapter that we must not in the mechanics of relativity necessarily assume the momentum to be in the direction of the velocity. If v_g is the component velocity in the direction of $\mathbf{g}$, remembering that $\mathbf{g}$ is perpendicular to $\mathbf{E}$ and $\mathbf{H}$, we have

$$c^2\,|\,\mathbf{g}\,| = v_g\,\{2w - |\,\mathbf{g}\,|\,v_g\},$$

or

$$2wv_g = |\,\mathbf{g}\,|\,(c^2 + v_g^2) \quad \text{.................(2)},$$

which, given w and $\mathbf{g}$, is an equation to find v_g.

4. The relativity of the equation for the velocity.

We may next examine whether the form of the condition (1) is maintained when we apply the transformations of the principle of relativity.

We have seen that on multiplying the 6-vector $\mathfrak{F} = (\mathbf{H}, -i\mathbf{E})$ by the 4-vector

$$\mathfrak{v} = \kappa\,(v_x, v_y, v_z, ic)/c,$$

we obtain a new 4-vector

$$[\mathfrak{v}\mathfrak{F}] = \kappa\,\{\mathbf{E} + [\mathbf{vH}]/c,\; -i\,(\mathbf{vE})/c\}.$$

Multiply $\mathfrak{F}$ again by this. Then

$$[[\mathfrak{v}\mathfrak{F}]\,\mathfrak{F}] = \kappa\,\{-\mathbf{E}\,(\mathbf{vE})/c + [\mathbf{EH}] + [[\mathbf{vH}]/c, \mathbf{H}],\; -i\mathbf{E}^2 - i\,(\mathbf{E}, [\mathbf{vH}]/c)\}$$

is a 4-vector.

Similarly

$$[[\mathfrak{v}\mathfrak{F}_1]\,\mathfrak{F}_1] = \kappa\,\{-\mathbf{H}\,(\mathbf{vH})/c + [\mathbf{EH}] + [[\mathbf{vE}]/c, \mathbf{E}],\; -i\mathbf{H}^2 + i\,(\mathbf{H}, [\mathbf{vE}]/c)\}$$

is a 4-vector.

Adding and multiplying by c we get another 4-vector, which on reduction becomes*

$$-2\kappa\,\{\mathbf{E}\,(\mathbf{vE}) + \mathbf{H}\,(\mathbf{vH}) + c^2\mathbf{g} - \mathbf{v}w,\; ic\,(w - (\mathbf{vg}))\}.$$

Now if the relation (1) is satisfied in any one frame of reference, this reduces to

$$-2\kappa\,(w - (\mathbf{vg}))\,\{\mathbf{v}, ic\},$$

that is, the given relation expresses that the 4-vector $[[\mathfrak{v}\mathfrak{F}]\,\mathfrak{F}]$ is in the same direction as the 4-vector $\mathfrak{v}$, that is, $\kappa\,\{\mathbf{v}, ic\}$. Hence if the relation (1) can be satisfied in any one set of coordinates it will also be satisfied in any other frame of reference. Further, the ratio of two 4-vectors in the same direction is an invariant; hence

$$w - (\mathbf{vg}) \text{ is invariant.}$$

* Using the identities

$$(\mathbf{E}\,[\mathbf{vH}]) \equiv (\mathbf{v}\,[\mathbf{EH}]) = (\mathbf{H}\,[\mathbf{Ev}])$$

and

$$[[\mathbf{vH}]\,\mathbf{H}] \equiv \mathbf{v}\,(\mathbf{H})^2 - \mathbf{H}\,(\mathbf{vH}).$$

5. The solution of the equation for the velocity.

It remains to be seen whether the equation (1) can be satisfied at all, and if so whether the solution is unique. To do this we may take any frame of reference we choose.

Now in general we can always find a frame of reference for which the electric and magnetic force at a point are in the same direction, so that in this system $\mathbf{g} = 0$.

For in an arbitrary field let the axis of x be taken perpendicular to $\mathbf{E}$ and $\mathbf{H}$, and apply the Lorentz transformation. Then

$$E_x' = E_x = 0, \quad H_x' = H_x = 0,$$

and
$$\frac{E_y'}{E_z'} = \frac{E_y - \frac{v}{c} H_z}{E_z + \frac{v}{c} H_y},$$

while
$$\frac{H_y'}{H_z'} = \frac{H_y + \frac{v}{c} E_z}{H_z - \frac{v}{c} E_y}.$$

Equating these we find in general that we can find two values of v for which $\mathbf{E}'$ and $\mathbf{H}'$ are in the same direction. The equation obtained is $ev\,(E_y H_z - E_z H_y) = (c^2 + v^2)(E^2 + H^2)$, which is exactly the equation obtained above for v_g.

Supposing then we so choose the frame of reference, the equation (1) becomes

$$\mathbf{E}'(\mathbf{v}'\mathbf{E}') + \mathbf{H}'(\mathbf{v}'\mathbf{H}') = 2\mathbf{v}'w',$$

so that, $\mathbf{E}'$ and $\mathbf{H}'$ being in the same direction, $\mathbf{v}'$ must also be in that direction and the equation then becomes an identity; thus the magnitude of $\mathbf{v}'$ becomes arbitrary. Transforming back to the original frame of reference, this arbitrary velocity along the direction of $\mathbf{E}'$ and $\mathbf{H}'$, say $(0, v_y', v_z')$, transforms into $(v_g, \beta v_y', \beta v_z')$, of which the first component has been shewn to be determinate in terms of $\mathbf{E}$ and $\mathbf{H}$ and the second and third give a velocity in a determinate direction in the plane of yz, that is, of $\mathbf{E}$ and $\mathbf{H}$.

Thus we find that *the reconciliation of the so-called stress with the assumed distribution of momentum and energy, and with the Poynting vector, determines the component velocity of the medium in the direction of the momentum, and the direction but not the magnitude of the remaining component.*

6. The total velocity at any point equal to that of light.

The relativity of the condition (1) shews that the determinate velocity v_g in a certain frame of reference, together with an arbitrary velocity in the determinate direction in the plane of $\mathbf{E}$ and $\mathbf{H}$ in that frame of reference, will by the kinematical transformation become the proper velocity v_g' together with another velocity in the appropriate direction in the plane of $\mathbf{E}'$ and $\mathbf{H}'$. We can remove the arbitrariness in the latter component by assigning a further condition to be satisfied by the velocity, provided that this condition conforms to the principle of relativity. Such a condition, and in fact the only such kinematical condition which we can lay down regardless of any particular circumstances, is *that the total velocity shall be equal to c.* We see therefore that it is possible to assign to the aether the velocity c in a proper direction at every point, in such a way that all the defining equations are consonant with the requirement of relativity, besides allowing of the maintenance of the principles of conservation of momentum and energy as between aether and the carriers of electric charge.

7. Two examples of the distribution of velocity.

(i) *A train of plane waves.*

In such a train of waves we know that the electric and magnetic intensities are equal and at right angles.

If $\mathbf{E}$ is the electric intensity, we have therefore

$$|\mathbf{g}| = \mathbf{E}^2/c, \quad w = \mathbf{E}^2,$$

and

$$\frac{2v_g c}{v_g^2 + c^2} = \frac{|\mathbf{g}|\, c}{w} = 1,$$

giving

$$v_g = c.$$

Thus in this case the whole velocity c is in the direction of propagation.

(ii) *The field due to a point charge moving in any manner.*

It is known that in this case, if $\mathbf{E}$ is the electric intensity at any point P at time t, and $\mathbf{r}_1$ is a unit-vector in the direction OP, where O is that point which was occupied by the point charge at a time t_1 such that $OP = c(t - t_1)$, then

$$\mathbf{H} = [\mathbf{r}_1 \mathbf{E}]^*.$$

Let PQ be the direction of $\mathbf{E}$; $\mathbf{H}$ is then perpendicular to the plane OPQ, and the direction PR of the momentum vector is in the plane OPQ and at right angles to PQ.

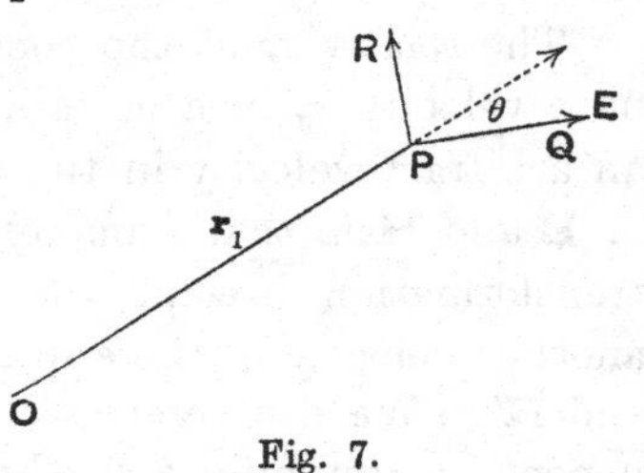

Fig. 7.

We have

$$|\mathbf{H}| = |\mathbf{E}| \sin\theta,$$
$$|\mathbf{g}| = |\mathbf{E}|^2 \sin\theta/c,$$

so that

$$\frac{2v_g c}{c^2 + v_g^2} = \frac{2\mathbf{E}^2 \sin\theta}{\mathbf{E}^2(1 + \sin^2\theta)},$$

giving

$$\frac{v_g}{c} = \sin\theta.$$

Thus the component velocity of the aether in the direction PR is $c \sin\theta$.

Multiply equation (1) by $\mathbf{H}$, remembering that

$$(\mathbf{gH}) = 0,$$

and

$$(\mathbf{EH}) = 0.$$

Then

$$\mathbf{H}^2(\mathbf{vH}) = (\mathbf{vH})\{2w - (\mathbf{vg})\},$$

or

$$\mathbf{E}^2 \sin^2\theta\,(\mathbf{vH}) = (\mathbf{vH})\{\mathbf{E}^2(1 + \sin^2\theta) - \mathbf{E}^2 \sin^2\theta\}$$
$$= (\mathbf{vH})\,\mathbf{E}^2.$$

Hence

$$(\mathbf{vH}) = 0,$$

that is, the remaining component of velocity is along PQ.

It follows that the total velocity c is along the direction OP. Thus the kinematical specification in this case is

* See Abraham, *Theorie der Elektrizität*, p. 96, 2nd edition, 1908.

simply that the aether travels as if emitted with velocity c from the moving charge, and any element of it moves always in a straight line with constant velocity.

In particular for a stationary charge the motion is a uniform flow outwards from the point with velocity c at all points.

8. The modified specification of stress in the moving aether.

Returning to the general case, for simplicity let the axis of x be taken in the direction of $\mathbf{g}$, so that

$$E_x = H_x = 0,$$

and

$$g_y = g_z = 0.$$

Further let the axis of y be taken in the direction of the remaining component of the velocity of the aether, so that $v_z = 0$.

Then the true stress tensor becomes

$$\begin{vmatrix} -w + v_x g_x, & v_y g_x, & 0 \\ 0 \quad , & Y_y \, , & 0 \\ 0 \quad , & 0 \, , & Z_z \end{vmatrix},$$

where

$$Y_y = -Z_z = \tfrac{1}{2}(E_y^2 + H_y^2 - E_z^2 - H_z^2).$$

Taking the y component of equation (1), we have

$$E_y(\mathbf{vE}) + H_y(\mathbf{vH}) = v_y(2w - (\mathbf{vg})),$$

which must be an identity so that, since

$$(\mathbf{vE}) = v_y E_y, \quad (\mathbf{vH}) = v_y H_y,$$

we have

$$E_y^2 + H_y^2 = 2w - (\mathbf{vg}),$$

or

$$(\mathbf{vg}) = E_z^2 + H_z^2.$$

Hence

$$\begin{aligned} -(X_x + v_x g_x) &= w - (\mathbf{vg}) \\ &= \tfrac{1}{2}(E_y^2 + H_y^2 - E_z^2 - H_z^2) \\ &= Y_y \\ &= -Z_z. \end{aligned}$$

If we use P to stand for the common value of these expressions, the stress tensor reduces to

$$\begin{vmatrix} -P, & v_y|\mathbf{g}|, & 0 \\ 0\ , & P\ , & 0 \\ 0\ , & 0\ , & -P \end{vmatrix},$$

which is the simplest form to which it may be reduced.

The form of this stress system is very similar to the simple Maxwell electrostatic system, the only difference being in the kinetic term $v_y|\mathbf{g}|$, a type of term which must always occur when the momentum is no longer restricted to be in the direction of the velocity. For the case of an electrostatic system $\mathbf{g} = 0$, and the stress becomes exactly the Maxwell stress.

But further, from the equation (2),

$$\frac{2v_g}{c^2 + v_g^2} = \frac{|\mathbf{g}|}{w},$$

we have
$$\frac{w - (\mathbf{vg})}{w} = \frac{c^2 - v_g^2}{c^2 + v_g^2},$$

so that
$$w = \frac{c^2 + v_g^2}{c^2 - v_g^2} P,$$

and
$$|\mathbf{g}| = \frac{2v_g}{c^2 - v_g^2} P.$$

Thus as far as this representation of the aether is concerned, it is completely defined if we are given

(i) The direction of $\mathbf{g}$.

(ii) The direction of the total velocity c.

(iii) The magnitude of what may be called the 'principal tension' P.

The equation (2) gives two possible values for v_g whose product is c^2. It is clear therefore that we must take always that root which is less than c.

Since
$$w - (\mathbf{vg}) \equiv \frac{c^2 - v_g^2}{c^2 + v_g^2} w,$$

and $v_g < c$, we shall have always, since w is positive,

$$w - (\mathbf{vg}) > 0.$$

Thus *the stress always consists of a 'tension' along the direction of v_y, that is in the direction of that component of the velocity of the aether which is perpendicular to the direction of the momentum, and of equal pressures in directions perpendicular to this, together with the kinetic terms.*

In the case of a field in which the electric and magnetic intensities are at right angles, this direction becomes that of the electric intensity, so that we arrive exactly at the conception of moving tubes of force in a state of stress of Maxwell's electrostatic type; the transverse velocity of the tubes is in fact the same as that proposed by Sir J. J. Thomson.

9. Examples of the distribution of stress.

In the examples given above the specifications of the stress are as follows:

(i) *The train of plane waves.*

Since $w = c\,|\,\mathbf{g}\,|$ and $v_g = c$,

$$w - (\mathbf{vg}) = 0.$$

Thus the principal stresses are all zero, and so is the term $v_y\,|\,\mathbf{g}\,|$, since $v_y = 0$.

Thus the train of plane waves is represented by a *purely convected momentum free from aethereal stress.*

(ii) *The field due to a moving point charge*

$$\begin{aligned} w - (\mathbf{vg}) &= \tfrac{1}{2}\mathbf{E}^2(1 + \sin^2\theta) - \mathbf{E}^2\sin^2\theta \\ &= \tfrac{1}{2}\mathbf{E}^2\cos^2\theta. \end{aligned}$$

Thus the stress tensor becomes

$$\tfrac{1}{2}\mathbf{E}^2\cos^2\theta \begin{vmatrix} -1, & \tan^2\theta, & 0 \\ 0, & 1, & 0 \\ 0, & 0, & -1 \end{vmatrix}.$$

The direction of the 'principal tension' is the direction of $\mathbf{E}$.

10. The 'stress' an absolute quantity.

At this point we may recall the remark made earlier (p. 6) that, in the Newtonian mechanics, while the 'velocity' of a body is a *relative* quantity, the 'force' acting on it, and connected with this the 'intensity of stress in a strained body,' are *invariant* quantities.

Now we have seen above that the quantity

$$P = w - (\mathbf{vg})$$

is an invariant, and this gives the intensity of the principal stress in the reduction that has been effected above.

Thus the kinematical properties of the suggested moving aether conform completely to the principle of relativity, while the *state of stress is invariant*; this is in complete analogy with the Newtonian conceptions.

11. The analysis of this chapter has been inserted with a view to shewing that an objective aether is not necessarily foreign to the point of view of the principle of relativity, and incidentally it appears that the mechanical categories of momentum, energy, and stress can also be maintained in their entirety. The only real modification to the ordinary theory of material media is that we can no longer take momentum to be in the direction of or proportional to the whole velocity.

If this view of the aether is accepted, the aether becomes much more nearly assimilated to our conception of an objective reality than on the older view where the relations expressed by the differential equations for the field were vaguely conceived as representing a quasi-elastic motionless medium, which served at the same time as a space frame of reference. At the same time the framework of time and space relative to which the aether moves in the presentation given above is restored to its true place as a mental construct embodying one aspect of the uniformities observed in the physical realm.

CHAPTER XVI

RELATIVITY AND PROBABILITY

1. General criterion of equally probable states.

In the kinetic theory of gases, and in the theory of radiation, one of the fundamental problems is to compare the relative probabilities of different states of a given system, of which not enough is known to determine the actual state. Various criteria of what may be taken to be equally probable states have been suggested, mostly based on what are conceived to be fundamental or intuitive conceptions.

For example if, in the kinetic theory of gases, nothing is known of a given particle save that it is within a certain region of space, it is assumed that all positions within that region are equally likely to be the actual position of the particle; or again it is assumed that for a given particle of which nothing is known to restrict its velocity, all velocities are equally probable, no matter how great.

Now in the case of the velocity it is obvious that the principle of relativity cannot admit this as a reasonable assumption, since the continual addition of velocities never leads to a velocity greater than that of light; and so the question may be asked: 'What criteria of equal probability are consistent with the principle?'

A general criterion may be laid down applying to all cases.

Any two states of a self-contained system which can be transformed into one another by a Lorentz transformation are to be considered as equally probable.

This clearly includes the first of the cases just cited, and agrees with the second provided we confine ourselves to small velocities.

2. The FitzGerald-Lorentz Contraction.

A first application of this criterion may be made to the theory of the FitzGerald-Lorentz contraction*. In respect of this the argument of Lorentz really seeks to establish nothing more than that, given any possible solution of the equations which determine the motions of the constituent elements of a body at rest, there exists another solution, correlated with the first by a certain transformation, corresponding to a uniform translation of the body as a whole.

If we adopt the criterion above mentioned, we can only say that, *ceteris paribus*, these correlated states are *equally probable*.

But the molar configuration of a material body, in particular what we call the length of the body, does not correspond to a unique solution of the molecular equations. We may take it to be assured, in spite of the incompleteness of the theory of the matter, that the molar properties of a solid body are statistical properties, and that there is a very great number of molecular configurations which are consistent with any given molar state, whereas there is only one molar configuration consistent with a given molecular state.

Thus we should say that the *natural* length of a body in given conditions is that which is statistically most probable.

But, since there is a one-one correspondence between the possible molecular configurations of a body at rest and those of a body in motion, it follows that the most probable molar configuration of a body at rest is that which is correlated with the most probable molar state of the body in motion.

This would explain to some extent why the FitzGerald contraction is automatic, if not instantaneous. If a body is set in motion, there will initially be set up in the body a disturbance of magnitude depending on the acceleration. This disturbance

* It is sometimes justifiably objected that the Einstein view of this contraction does not make it clear that when a body is accelerated relative to a given observer, it will actually take up in the definite frame of reference of that observer the contracted form which would be apparent if the body remained unaltered in its motion and the frame of reference changed.

will be gradually dissipated by radiation or internal friction, the body gradually settling down into the most probable state consistent with its altered velocity. To any observer moving with the body in its new motion this state will be the same as the original state of the body, and therefore to any other observer the body will appear contracted according to the Einstein transformation, exactly as FitzGerald suggested.

3. **Examples from the Kinetic Theory of Gases.**

We may next give one or two examples of the modifications required to the ordinary results of the kinetic theory of gases.

The probability of the velocity of a particular molecule in a given mass of gas lying between v and $v + \delta v$ is ordinarily taken to be of the form $Ae^{-av^2}\delta v$, a being a constant. This is based on the criterion of *a priori* equal probability for all velocities.

With the criterion above described the result is as follows: *If nothing is known of a given large number of moving particles save that they have a given constant total energy, the energy 'w' of each being a given (the same for all) function of its velocity, then the probability of the velocity of a particular particle lying between v and $v + \delta v$ is of the form*

$$Ae^{-aw}v^2\left(1-\frac{v^2}{c^2}\right)^{-2}\delta v.$$

If it be assumed that the particles are all identical to observers moving with them, w is known as a function of v. It is, as was shewn above, equal to $w_0(1 - v^2/c^2)^{-\frac{1}{2}}$, where w_0 is a given constant, the energy of a particle at rest. If v/c is small it is seen at once that this reduces to the customary expression.

4. The proof of the above expression is as follows.

Let us compare a particle with small velocity (u_x, u_y, u_z) with one obtained from it by the transformation of Einstein.

Its velocity is given by

$$u_x' = \frac{u_x + v}{1 + \frac{vu_x}{c^2}} = v + u_x\left(1 - \frac{v^2}{c^2}\right) = v + u_x/\beta^2,$$

and $$u_y' = u_y/\beta, \quad u_z' = u_z/\beta,$$

neglecting $(u_x/c)^2$, and the products $u_x u_y/c^2$ and $u_x u_z/c^2$. Hence if, in a velocity diagram, (u_x, u_y, u_z) is represented by a point within a small element of volume $\delta\tau$ about a point near the origin, the velocity (u_x', u_y', u_z') is represented by a point within an element of volume $\delta\tau'$ about a point near to $(v, 0, 0)$, where

$$\delta\tau' = \delta\tau/\beta^4.$$

Our criterion of probability states that *it is just as probable that the velocity of any particular particle shall be represented by a point within* $\delta\tau$ *as within* $\delta\tau'$.

Now let us divide the whole region bounded by the sphere of radius c in the velocity diagram into small cells of equal probability, the volume of any one at distance v from the centre being therefore proportional to β^{-4}, say equals $\tau\beta^{-4}$. Calling the numbers of particles whose velocities are represented by points within the respective cells $n_1, n_2, n_3 \ldots$ and the energy corresponding to the same cells $w_1, w_2, w_3 \ldots$, the probability Π of a given distribution is proportional to

$$\frac{\underline{|N}}{\underline{|n_1}\,\underline{|n_2}\,\underline{|n_3}\ldots},$$

where N is the total number of particles.

The total energy is

$$n_1 w_1 + n_2 w_2 + n_3 w_3 + \ldots,$$

which is a given quantity.

If now we seek to make Π a maximum subject to the given conditions, we have in the ordinary way

$$n_r = Ae^{-\alpha w_r},$$

A and α being constants*.

In order to find the number of particles whose resultant velocity lies between v and $v + \delta v$, we must multiply n_r by the number of cells contained in a spherical shell of radius v and thickness δv, the volume of each cell being $\tau\beta^{-4}$.

* See, for instance, Jeans, *Dynamical Theory of Gases*, pp. 39 ff.

Thus the required number is

$$\frac{A \,.\, 4\pi v^2 \delta v e^{-aw}}{\tau \beta^{-4}} = C\beta^4 v^2 e^{-aw}\, dv,$$

shewing that the probability is as stated above.

The probability of the velocity of a single particle lying between limits (u_x, u_y, u_z) and $(u_x + du_x,\ u_y + du_y,\ u_z + du_z)$ is similarly found to be proportional to

$$\beta^4 e^{-aw}\, du_x du_y du_z.$$

5. These results are obtained simply on the assumption of given total energy for the particles. If however we are considering only those particles which lie in a given volume, there is a further point to be considered. Owing to the difference in the measure of a volume according as it is conceived to be at rest or in motion, the criterion of equal probability tells us that other things being equal the probability of a particle of velocity v lying within a volume $\delta\tau/\beta = \delta\tau\,(1 - v^2/c^2)^{\frac{1}{2}}$ is the same for all values of v. This has the effect of increasing the probability of the velocity v in a ratio proportional to β, so that the *number of particles in a given volume* having velocities between the limits named becomes proportional to

$$\beta^5 e^{-aw}\, du_x du_y du_z {}^{*}.$$

6. Illustration from the theory of radiation.

In his Bakerian Lecture to the Royal Society (1909), Larmor introduces the idea of equally probable elements of radiation, or equally probable trains of waves, using as the criterion of equal probability of two elements, the possibility of a physically reversible transformation of the one into the other, as for example by means of reflection at a moving mirror. In this way two groups of equal numbers of plane waves of the same area but of different wave-lengths are counted as equally probable if the total energies in the groups are proportional to the frequencies, these frequencies not being given exactly but

* This result is obtained by a different method by Jüttner, *Ann. der Physik*, 34 (1911), p. 857.

being known to lie within two small limits which are proportional to the respective frequencies.

Let us now apply the criterion of relativity for the comparison of waves of different frequencies instead of the criterion suggested by Larmor.

Suppose that we consider a train of plane waves of period τ travelling parallel to the axis of x. If we apply a Lorentz transformation to such a group of waves, we have, since for the plane waves

$$E_x = H_x = 0, \quad E_y = H_z, \quad E_z = -H_y,$$

$$\begin{aligned} W' &= \tfrac{1}{2}(E_y'^2 + H_y'^2 + E_z'^2 + H_z'^2) \\ &= \tfrac{1}{2}\beta^2 \left\{\left(E_y - \frac{v}{c}H_z\right)^2 + \left(H_y + \frac{v}{c}E_z\right)^2 + \left(E_z + \frac{v}{c}H_y\right)^2 \right. \\ &\qquad\qquad \left. + \left(H_z - \frac{v}{c}E_y\right)^2\right\} \\ &= \tfrac{1}{2}\beta^2 \left\{2E_y^2\left(1 - \frac{v}{c}\right)^2 + 2H_y^2\left(1 - \frac{v}{c}\right)^2\right\} \\ &= \beta^2\left(1 - \frac{v}{c}\right)^2 W, \end{aligned}$$

W, W' being the energy densities of the groups of waves as seen in the two frames of reference.

But as we have seen above*, the equation connecting the respective periods or wave-lengths for the two systems is

$$\frac{\lambda'}{\lambda} = \frac{\tau'}{\tau} = \frac{1}{\beta\left(1 - \frac{v}{c}\right)}.$$

Thus, since the corresponding volumes occupied by the groups of waves are in the ratio $\lambda' : \lambda$, the *total* energies E', E in the two groups are such that

$$\frac{E'}{E} = \beta\left(1 - \frac{v}{c}\right) = \frac{\tau}{\tau'} = \frac{\nu'}{\nu},$$

where ν' and ν are the frequencies.

* Chap. VI, p. 63.

Also, since for a Lorentz transformation corresponding to a velocity v the ratio $\frac{\nu'}{\nu} = \beta\left(1 - \frac{v}{c}\right)$ is independent of ν, if we consider a range of frequency ν to $\nu + \delta\nu$ the corresponding range in the transformed system will be from ν' to $\nu' + \delta\nu'$, where

$$\frac{\nu'}{\nu} = \frac{\nu' + \delta\nu'}{\nu + \delta\nu} = \frac{\delta\nu'}{\delta\nu}.$$

Thus if we adopt the relativity criterion of equal probability we conclude that a group of plane waves of total energy E, whose frequency lies between two limits ν and $\nu + \delta\nu$, is equally probable with a group of the same number of waves of total energy E' and frequency lying between ν' and $\nu' + \delta\nu'$, the areas of the wave fronts being identical, provided that

$$\frac{E'}{E} = \frac{\nu'}{\nu} = \frac{\delta\nu'}{\delta\nu},$$

a resultant exactly according with that obtained by Larmor by a consideration of quite a different nature.

This argument is in accord with the now well-known hypothesis that light energy exists in discrete quanta of different frequencies, the amount of energy per quantum being proportional to the frequency. All such quanta would from the present point of view be 'equally probable.'

7. **Relativity and Thermodynamics.**

The foregoing results obtained from the suggested criterion of probability suggest at once that if we are to adopt the principle of relativity in its entirety, not only will dynamical results have to be modified, but there will also have to be a reconsideration of the definitions and properties of thermodynamic quantities. Such magnitudes as temperature will no longer be defined in a manner which is completely independent of the kinematical state of the system. The temperature of a system will only have a definite value when the velocity of the system is specified.

The most general way of approaching this aspect of the principle is through the definition of the 'entropy.' If we adopt Boltzmann's definition, in which the entropy is the logarithm of the probability of a given state of the system, it follows at once from the general criterion of equal probability that the *entropy is an invariant.*

We have also seen above that the *pressure* is an *invariant.* Having gone so far it is easy to see what must be postulated of the *temperature* in order to maintain complete relativity. If w_0 is the energy of a system at rest, and w of the same system supposed to have a velocity v, we have

$$w = w_0 (1 - v^2/c^2)^{-\frac{1}{2}}.$$

Thus, if the energy of the system is changed,

$$dw = dw_0 (1 - v^2/c^2)^{-\frac{1}{2}},$$

and since the entropy S is invariant

$$dS = dS_0,$$

so that

$$\frac{dS}{dw} = (1 - v^2/c^2)^{\frac{1}{2}} \frac{dS_0}{dw_0}.$$

If therefore we desire to develop a scheme of thermodynamics in which the relation

$$\frac{dS}{dw} = T$$

holds independently of the velocity attributed to the moving system, we must assume that the temperature is subject to the transformation

$$T = T_0 (1 - v^2/c^2)^{\frac{1}{2}},$$

or

$$T/\beta \textit{ is invariant.}$$

The above relations were first obtained by Planck from a different point of view. For a further development of the relations between the thermodynamics of a moving system and its motion, reference may be made to his work*.

* M. Planck, *Ann. der Phys.* 26 (1908), p. 1. See also Laue, *Das Relativitätsprinzip*, pp. 212–223.

CHAPTER XVII

CONCLUDING REMARKS

1. The main objections urged against the Principle of Relativity are (i) that it is unnecessary and too sweeping, (ii) that it does away with the possibility of an objective aether, and (iii) that time and space are such immediate objects of perception that the artificial view which it adopts of them cannot in any sense correspond to reality.

2. In respect of the last difficulty little can be said to meet the natural shrinking which the observer of natural phenomena feels from such a calculus as Minkowski's, in which we seem to lose sight of the most obvious distinction between time and space as essentially different modes of ordering events.

It must be remarked, however, that an essential part in the practice of the calculus is the final process of interpreting the analytical result in terms of the ordinary modes of thought. There is perhaps an analogy to be drawn between the analysis which lays out the whole history of phenomena as a single whole, and the things in themselves, the natural phenomena apart from the human intelligence, for which consciousness of time and space does not exist, the laws of which, when expressed for instance by means of a principle of least action, consist in a relation between the whole aggregate of configurations which their history contains; in which, so far as they are mechanically determinate, the past and the future are interchangeable. Such a view of the universe is inseparable from a mechanical determinism in which the future is unalterably determined by the past and in which the past can be uniquely inferred from the present state of the universe. It is the view of an intelligence which could comprehend at one glance the whole of time and space.

But the limitations of the human mind resolve this changeless whole into its temporal and spatial aspects, and the past and future of the physical world is the past and future of the intelligence perceiving it. Only to a being outside the physical universe, free from participation in its phenomena, is time a meaningless term. The human consciousness and the physical universe are inseparably parts of a greater whole. They run parallel to one another, and the brain cannot do otherwise than order physical and external events relative to the internal sequences of its own consciousness.

It is by such a process of correlation that any analytical scheme of relations is constructed for the description of natural processes. When this has been carried out, it is claimed for it that it, at any rate approximately, contains within it the whole history of those processes for the mind to grasp as one whole. Thus the very act of formulating a set of equations which make the present state of the system to contain implicitly within it the whole history, past and to be, is one step, and that the largest, towards eliminating the peculiar characteristic of time as a product of the inner consciousness from its place in physical relations. It is but a small step further to the timeless universe of Minkowski.

It is in fact the sole aim of theoretical physics to distinguish between and disentangle one from the other those factors in perceived events which are dependent upon human consciousness and those which are completely independent of it. The achievements of the past in this direction are quite sufficient to warrant further and continuous effort. That the mind should be able to conceive such a daring project and to progressively realize it, seems almost in itself sufficient to indicate that the resolution of its own workings into a chain of physically determinate processes is one incapable of complete realization.

3. The other objections to the principle of relativity raised have been dealt with in the course of this work. It has been shewn in Chapter XV that the principle of relativity is not inconsistent with the conception of an aethereal medium as the carrier of electromagnetic effects, though it is inconsistent with the unnecessarily restricted rigid aether which has been commonly presented. It is hardly hoped that the suggestion as to the way in which the reconciliation may be effected will be at all commonly accepted at present, but at any rate the force of the objection, such as it is, on this score is diminished.

As a matter of fact, the trend of thought among certain physicists at the present time is rather towards the assertion that the classical conception of the aether is insufficient to comprehend certain new experimental facts that have emerged, especially in the matter of radiation, photo-electric effects, specific heat. There is a tendency to think of radiation as having a more definite individuality than that of a mere disturbance in a continuous medium. The theory of energy-quanta is inconsistent with the usual view put forward as to the nature of radiation, and is suggestive of a modified emission theory of light. Such a view of the aether as has been suggested may perhaps tend to make the possibility of a reconciliation seem a little less remote. In any case it is free from the objection that it is confusing conceptual space with a conceptual quasi-material medium.

In restoring time and space to their lawful position as factors in human consciousness, the Principle of Relativity is preserving the true nature of physical theory as the description of facts in their mutual perceptual relations, and not in relation to a metaphysical background. There is really a greater logical justification for the efforts of those who have striven to correlate electrodynamics and mechanics by the construction of a mechanical aether than there is for giving to a frame of reference the status of an objective medium. To reduce the description of aethereal phenomena to a scheme of differential equations different in form from those characteristic of any type of material medium is really to do away with the aether as an effective means of visualizing the means of transmission of electromagnetic effects, especially when we remember that the only phenomena which we can perceive are localized in matter and not in the aether, so that the variables involved in the equations are not quantities capable of measurement except at points occupied by matter (or charge).

In fact the demand for an objective aether has no meaning until we have defined in what sense 'objectivity' is to be understood. A pure product of analysis is subjective if anything is. Objectivity for a conceptual medium can mean nothing more than similarity in some of the properties in terms of which it is defined to the observed properties of some perceptual medium. The interpretation of the properties of the aether in terms of some mechanical model has been tried by many, but no really satisfactory result has emerged, and the tendency has been to settle down to a meaningless statement that the aether is at rest, without in any way correlating that statement with any of its other properties. Such an aether is the most subjective of possible conceptions.

4. If the hypothesis of complete relativity may seem to some to have been rashly put forward and hastily adopted without an overwhelming evidence compelling it, it is perhaps not irrelevant to remark that nearly all the greatest generalizations of physical science have been made from a few isolated facts, but that their universal acceptance at the present day is the result of the unbroken agreement with facts since accumulated. To many present-day physicists Newton's enunciation of the Law of Gravitation would have seemed to be based on very scanty evidence, but it has been tested as it would seem to the utmost by the advance of astronomical science, and not been found wanting except in the particular instance of the unexplained motion of the perihelion of the orbit of Mercury; and even in this case it has been suggested that the defect is not in the law, but in our ignorance of certain hitherto unsuspected attracting bodies (Seeliger). The only objection felt to it at the present moment is that gravitation obstinately refuses to fit in with our slight knowledge of the constitution of matter.

At the present day we are only on the threshold of

investigating second order phenomena such as those anticipated by Michelson and Morley, Rayleigh and Brace, and it may be that for many years the experimental evidence accumulated will be insufficient to obtain for the Principle of Relativity, on the one hand a secure place, or on the other hand a relegation to the collection of outworn theories. At the present moment it is irrelevant to much of our experimental knowledge, but as far as we can tell it is consistent with it all, and it gives a unity to our thought which is not obtained by shewing that the few definite results of experiment can each be explained in some more conservative manner; and this especially in view of the gaps, some of which have been indicated, in existing theories.

5. In spite of all the elaborate structure of mathematical developments which has been built up in connection with the principle, the most important aspect of it is still that which, with a flash of insight, was first enunciated by Einstein, the assertion that metrical space and time are not independent and self-contained concepts, but that they are conditioned by the very phenomena which they are used to describe. Though now to many this seems an obvious truth, and though it has been implicit in the way in which many thinkers have always looked upon the meaning of all physical measurements, yet the point of view is still far from being generally adopted.

At present it is difficult to see whether the results forecasted by the principle are likely to have any considerable bearing on the growth of our experimental knowledge of the universe. It seems that we are entering on a new region of phenomena of untold possibilities for our insight into the constitution of matter. Much more must be done before so broad a generalization can be made as seemed only a few years ago possible in the conception of a matter built up of simple electrons*. But as this possibility becomes the more remote, the more importance attaches to the Principle of Relativity as an independent hypothesis; for the experiments which set it on foot can never be undone.

* Cf. Soddy, *Nature*, Nov. 20, 1913, p. 339.

INDEX

SELECTION FROM THE GENERAL CATALOGUE

OF BOOKS PUBLISHED BY

THE CAMBRIDGE UNIVERSITY PRESS

Aether and Matter: a development of the Dynamical Relations of the Aether to Material Systems on the basis of the Atomic Constitution of Matter, including a discussion of the influence of the Earth's motion on optical phenomena. Being an Adams Prize Essay in the University of Cambridge. By Sir JOSEPH LARMOR, D.Sc., LL.D., Fellow of St John's College and Lucasian Professor of Mathematics. Demy 8vo. 10s net.

An Elementary Treatise on the Dynamics of a Particle and of Rigid Bodies. By S. L. LONEY, M.A., Professor of Mathematics at the Royal Holloway College (University of London). Demy 8vo. 12s.

A Treatise on the Analytical Dynamics of Particles and Rigid Bodies; with an Introduction to the Problem of Three Bodies. By E. T. WHITTAKER, M.A. Royal 8vo. 12s 6d net.

Dynamics. By HORACE LAMB, LL.D., D.Sc., F.R.S. Demy 8vo. 10s 6d net.

Some Problems of Geodynamics. Being an essay to which the Adams Prize in the University of Cambridge was adjudged in 1911. By A. E. H. LOVE, M.A., D.Sc., F.R.S. Large Royal 8vo. 12s net.

The Dynamical Theory of Gases. By J. H. JEANS, M.A. Royal 8vo. 15s net.

A Treatise on the Kinetic Theory of Gases. By S. H. BURBURY, M.A., F.R.S. Demy 8vo. 8s.

A Treatise on Elementary Thermodynamics. By J. PARKER, M.A. Crown 8vo. 9s.

The Laws of Thermodynamics. By W. H. MACAULAY, M.A. Demy 8vo. 3s net.

The Thermal Measurement of Energy. Lectures delivered at the Philosophical Hall, Leeds. By E. H. GRIFFITHS, Sc.D., F.R.S., Fellow of Sidney Sussex College. Crown 8vo. 2s.

Baltimore Lectures on Molecular Dynamics and the Wave Theory of Light, followed by Twelve Appendices on Allied Subjects. By Lord KELVIN, O.M., G.C.V.O., P.C., F.R.S., &c. Demy 8vo. 15s net.

Radioactive Substances and their Radiations. By Sir ERNEST RUTHERFORD, D.Sc., Ph.D., LL.D., F.R.S., Nobel Laureate, Professor of Physics, Victoria University of Manchester. Demy 8vo. 15s net.

Electromagnetic Radiation and the Mechanical Reactions arising from it. Being an Adams Prize Essay in the University of Cambridge. By G. A. SCHOTT, B.A., D.Sc., Professor of Applied Mathematics in the University College of Wales, Aberystwyth. Large Royal 8vo. 18s net.

[*continued overleaf*